Experimental Fatigue Life Estimation of OWC Using Accelerated Life Testing

Vinod Mehra

CONTENTS

Chapter 1

Introduction

This chapter presents one-way clutch (OWC): basic theory and operation with their industrial applications. The geometrical and material parameters for OWC are discussed. The contact fatigue and wear observed for OWC are described. Motivation for the current research work is discussed along with the scope of work.

One-way power transmission elements, generally known as mechanical one-way clutch (OWC) or free-wheels, are used as backstopping devices in heavy power industries to safeguard the machines running with prime movers and gearboxes. The clutch transmits the torque in one direction and acts as a freewheel in the reverse.

1.1. One-way clutch

State-of-art one-way clutches are highly engineered mechanical products originally designed for passenger car and helicopter transmission systems of the automotive industries to automatically shift the torque converters into fluid couplings, to act as freewheels in constant-mesh spur gear transmission systems and to ease the shift-time problems in planetary type of automatic transmission systems **[1]**. Apart from that, OWCs are used in starter systems of turbojet engines to isolate the main engine from turbine rotors running at high speeds **[2]**. Further improvements in the automotive design of the clutch make it useful in critical applications such as helicopter transmission units to continue rotating the main rotor faster than the engine in the event of engine failure **[3]**. Over the years, OWC found its application in starter motors of automobile engines and modern motorcycles as primary drives between the engine and transmission systems to facilitate the backshifts **[4]**. In the late 90s, OWC made its appearance in mechanical power transmission industries, where it was utilized as a critical component to safeguard machinery running with prime movers and gearboxes **[5]**.

Initially, OWCs were considered as torque transmission or freewheel components as per the system's suitability, but the continued development towards increasing performance and reliability of such critical components has significantly reduced the occurrence of catastrophic failures. As a result, it opens up the opportunity for the manufacturing industries to promote more reliable and cost-effective solutions for one-way power transmission elements, thereby replacing conventional technologies such as pawl and ratchet-type devices. OWCs are classified based on their geometrical construction and suitability to various applications.

1.1.1 One-way clutch: types and applications

One-way power transmitting elements, or OWCs, have emerged as one of the most reliable products in power transmission industries. They are frequently employed in heavy industries, such as inclined belt conveyor systems, vertical turbine pumps, bucket elevators, etc., to safeguard the power source and other accessories. OWCs consist of outer race, inner race and contact elements incorporated between them as primary components. The load transfer between these elements is accomplished through mechanical contact by providing wedging action amongst them. In industries, various configurations of OWC are available based on their distinct designs and applications. The geometrical shape of the contact elements defines the type of configuration that is needed for the application. Some of them, which are relevant to current investigations, are discussed below.

1. Roller ramps or roller OWC
2. Sprag or cam clutch
3. Centrifugal lift-off sprag (CLS) clutch

In roller ramps or roller OWC, the inner race is provided with an accurate geometric shape to form a wedge pocket into which cylindrical rollers are operated through spring-operated plungers. The spring-energized plungers keep rollers in contact with races while transmitting torque. During the engaged mode of operation, the rollers positively lock the inner race, enabling the torque transmission to the outer race or vice-versa when rotated in one particular direction. Highly localized contact stresses are generated during this mode due to wedging action at roller-race contacts. On the other hand, the freewheel mode of operation makes the unit totally free by relative motions between inner-outer races when one race rotates in the opposite direction. The assembly parts for roller OWC are illustrated in **Figure 1.1 [6],** along with a sectional view of the components. As roller OWC is non-self-supporting, it requires additional bearing supports to exhibit perfect concentric operation with inner and outer races and eliminate axial loads due to possible misalignments.

Applications: Low-speed applications like conveyor industries, bucket elevator industries, etc.

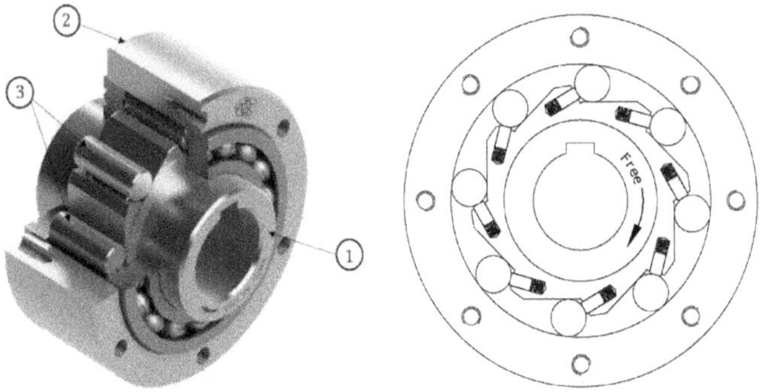

Figure 1.1 Roller one-way clutch assembly: (1) inner race, (2) outer race, and (3) cylinder rollers with sectional view on the right side. (Courtesy: NMTG Mechtrans Techniques Private Limited) **[6]**

Sprag-type OWC utilizes precisely formed cam elements, generally known as sprags energized with retainer springs between two races. Such assembly makes the clutch a freewheel OWC, as the rotation of any one race in one particular direction frees sprags around their rotational axis. It is because of the specific design of sprag geometry at the inner and outer contact regions of the races. The contact point changes when the clutch rotates in the reverse direction due to its eccentric design and accomplishes the wedge contact to positively lock the other race to further transmit the torque. The assembly parts of the sprag clutch are shown in **Figure 1.2 (a) [6]** with the sectional view.

Apart from the conventional design of cam elements, sprag-type OWCs are also classified based on the specific cam profiles and energizing springs, one of which is displayed in **Figure 1.2 (b) [6]** that makes it unique configuration utilized for various operating parameters that include size, space, operating speeds and variable loading conditions. The clutch with a retainer cage is provided with a torsional spring instead of a conventional coil spring to facilitate and operate sprags within the permissible angular limits.

Applications: Medium to high-speed applications like gear-box transmission systems, industrial fan systems- induced/forced draft fans, dual-drive systems like air preheater and ball mill drive, etc.

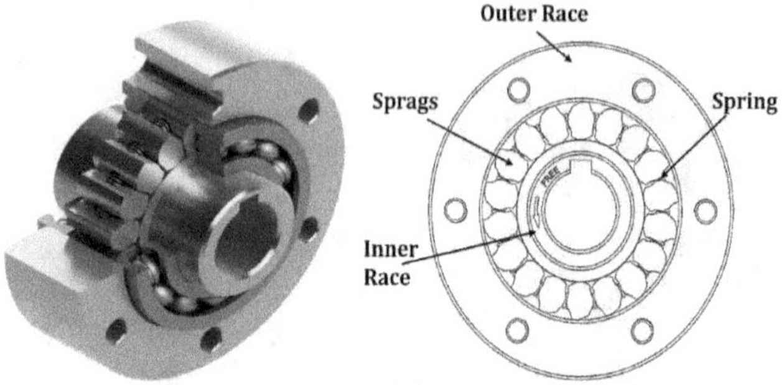

Figure 1.2 (a). Sprag-type one-way clutch assembly with sectional view on the right side. (Courtesy: NMTG Mechtrans Techniques Private Limited) **[6]**

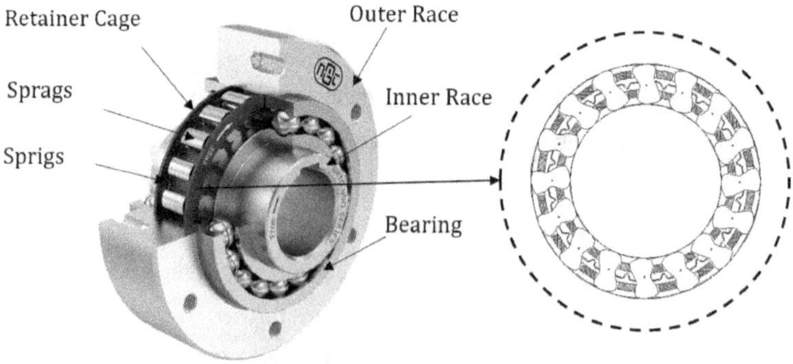

Figure 1.2 (b). Sprag-type one-way clutch assembly with retainer cage with sectional view on the right side. (Courtesy: NMTG Mechtrans Techniques Private Limited) **[6]**

Centrifugal lift-off type sprag (CLS) OWC is typically used as a contact-free one-way power transmission device, which works on the principle of centrifugal action of contacting elements. CLS-type OWC finds its application where there is an operating speed limit beyond which the clutch needs to function entirely free without any contact with the races. This specific function of the clutch can be achieved by locating the contact elements eccentrically with the help of torsion springs in cage plate assembly. These plates have specific counters that can restrict the sprag's angular movements within specific positional limits. The cage plate assembly is mounted on either of the races by press fitting. With this arrangement, sprags lift themselves off about the eccentric centre due to centrifugal force when any one of the races rotates in one particular direction with increasing speeds. This arrangement makes this clutch specifically worthwhile when there is a rotating speed limit beyond which it is necessary to operate the system under complete freewheel mode. On the other side, when the speed of the clutch falls below the lift-off speed or the clutch is rotated in the reverse direction, sprags tend to make driving contact with the races due to reactive spring force and eventually, the clutch initiates to transmit the torque. The CLS-type OWC and assembly parts are illustrated in **Figure 1.3 [6]** with the sectional view.

Application: High-speed applications like vertical turbine pumps, servo and induction motors, gearbox transmission systems, induced/forced draft fans, etc.

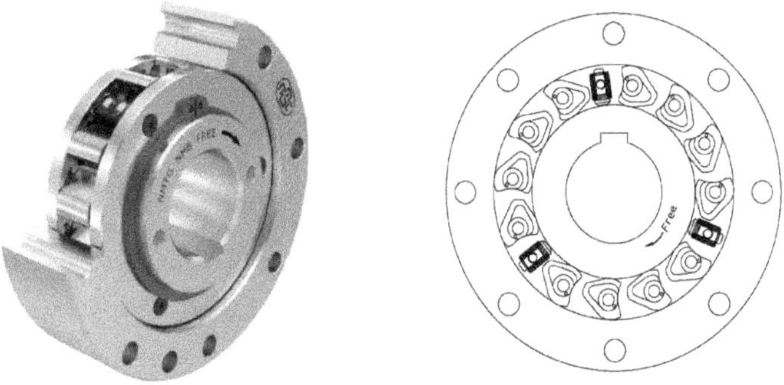

Figure 1.3. CLS-type one-way clutch assembly with a sectional view on the right side. (Courtesy: NMTG Mechtrans Techniques Private Limited) **[6]**

Operating Functions

OWCs are used for three basic transmission functions, which allow them to be utilized for three distinct modes of operation.

(a) Backstopping (holdback)

(b) Indexing (feeding)

(c) Overrunning (free-wheeling)

OWC, as a backstopping device, provides a fail-safe design to prevent reverse rotation of the load when the input drive fails to continue motion in case of power failure, thereby avoiding catastrophic failures **[7].** It could damage the entire system or machinery if OWC is not installed at the drive or non-drive end. The backstopping function can be achieved by fixing one of the clutch races to support the housing/bracket or any stationary frame through the torque arm/additional flange to restrict the reverse rotation to avoid damage to the supply unit.

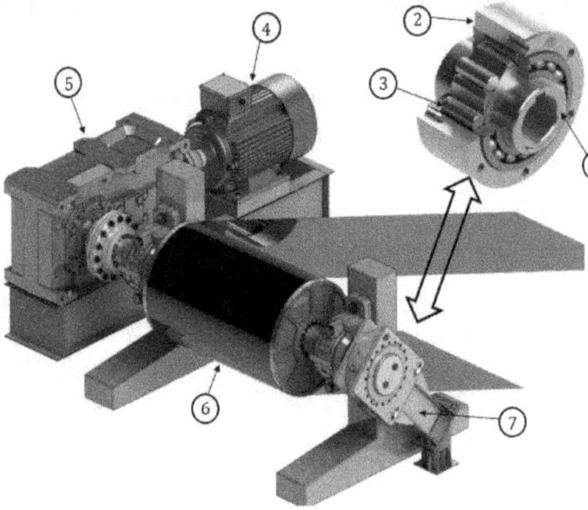

Figure 1.4 (a). One-way clutch in typical inclined belt conveyor application. (1. inner race, 2. outer race, 3. rollers, 4. motor, 5. gearbox, 6. conveyor pulley, 7. lever arm) (Courtesy: NMTG Mechtrans Techniques Private Limited) **[6]**

An inclined belt conveyor is an application where OWC is utilized as a backstopping device, as shown in **Figure 1.4 (a) [6]**. During the usual conveyor application, many other elements like gears, motors, and bearings operate continuously as per their functional requirements. OWC, on the other end, remains inactive during most of its operating time with the usual overrunning of races and only comes into action when a load must be stopped immediately after the drive has stopped/failed to prevent reverse rotation.

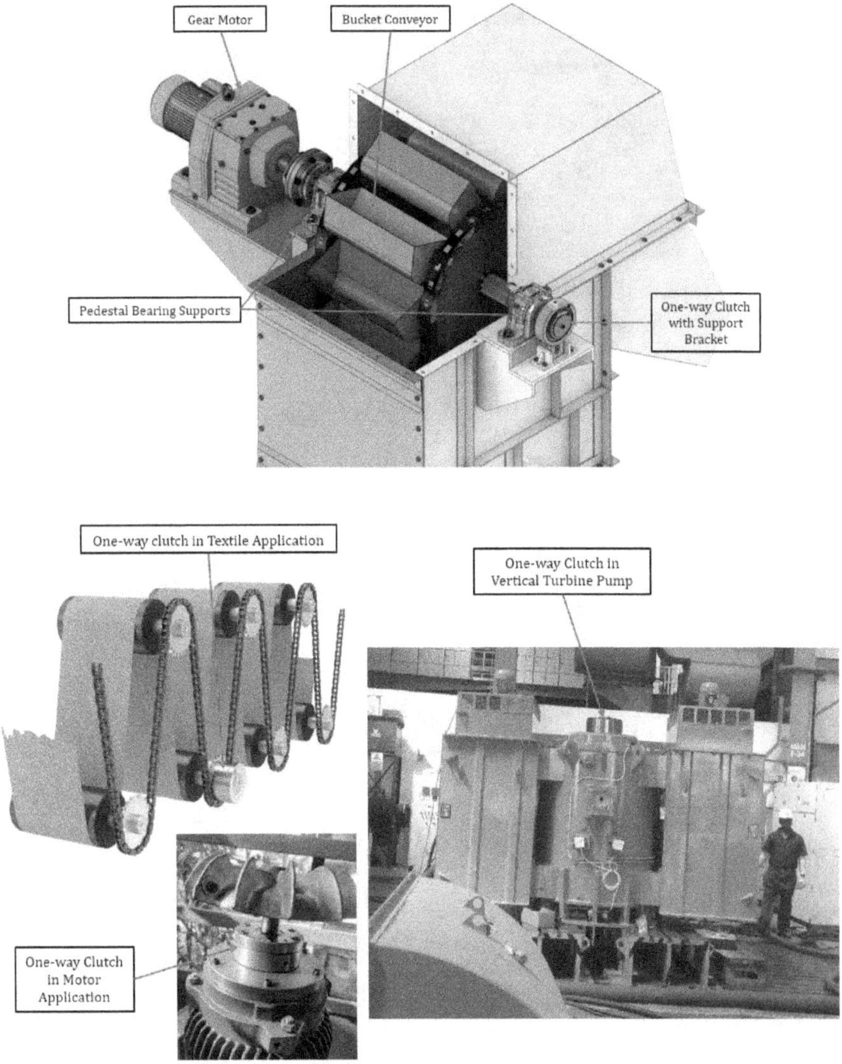

Figure 1.4 (b). One-way clutch in various industrial applications. [Source: through NMTG Mechtrans Techniques Private Limited] **[6]**

Figure 1.5 (a). OWC as overrunning clutch in the dual-drive application. (Courtesy: NMTG Mechtrans Techniques Private Limited) **[6]**

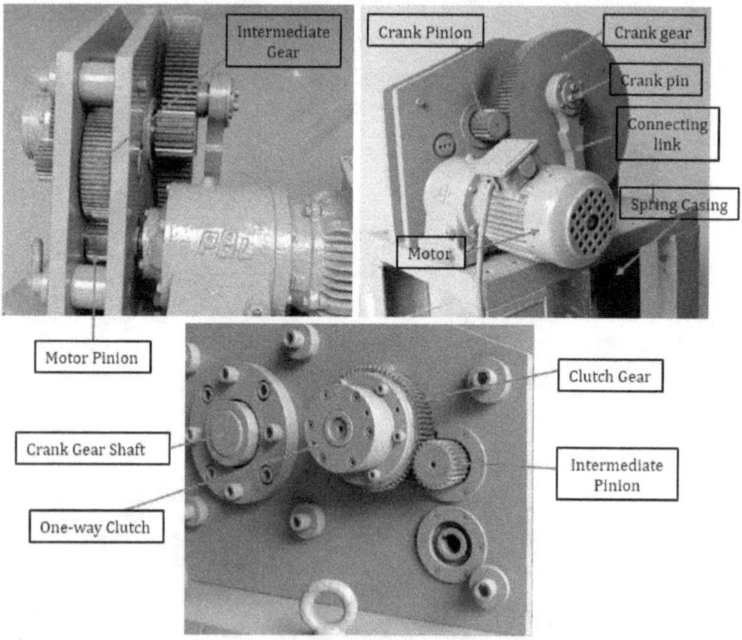

Figure 1.5 (b). OWC as overrunning clutch in the switch gear application. (Courtesy: NMTG Mechtrans Techniques Private Limited) **[6]**#

#Karan A. Dutt*, Shashikant J. Joshi, Dhaval B. Shah, Shashikant B. Soni, Deepak Prajapati. "Design and Development of Experimental Test Rig for Fatigue Life Estimation of Freewheel One-way Clutch in Switchgear Application at High Overrunning Speed." International Conference. ICMME 2023 organized by GLA University-Mathura.*

Further, the most general system arrangements employed by OWC, such as bucket conveyors, vertical turbine pumps, industrial motors, and textile applications, are displayed in **Figure 1.4 (b) [6]**. As an indexing device, OWC transmits the unidirectional torque in one direction while running free in reverse, thereby enabling the conversion of rotary motion to one race into the other's intermittent linear motion or vice versa.

OWCs are also used in the automotive and heay industries as mechanical overrunning clutches, mostly in dual-drive applications, as illustrated in **Figure 1.5 (a) [6],** where high start-up torque is required to the system, which cannot be initially accessed through the main drive. The clutch mostly overruns during normal operating conditions when the auxiliary drive gets disconnected automatically, and the main drive operates at normal operating speeds. In such applications, the clutch automatically disengages when the driven race rotates faster than the driving race. Apart from this, OWC is also employed in the spring-operated circuit breaker mechanism, as shown in **Figure 1.5 (b) [6],** to trip the electrical circuit under a high overload current. The application of OWC in the switchgear tripping mechanism utilizes the clutch's overrunning function to discharge the high-speed rotation of the driven unit when the circuit trips without affecting the drive unit. The basic theoretical principles on which OWC works is described along with the geometrical and material parameters in the next subsections.

1.1.2 One-way clutch: theory and operation

Modern OWCs provide simpler and easier installation than conventional ratchet-type devices **[8].** The wedging elements of the clutch are accompanied by energizing springs to keep the elements in contact with the races and achieve a one-way clamping function due to definite contact geometries. The elements are distinctly designed and provided with eccentric geometries to display the required clamping function in one particular direction. There are two modes of operation for traditional OWC: (i) engaged mode and (ii) freewheel mode. During the engaged condition, sprags positively lock the inner race with respect to the outer race, enabling torque transmission from the inner race to the outer race or vice versa when rotated in one specific direction. During this mode, the contact force is developed at the contact regions due to wedge action created, which tends to generate contact stresses at sprag-race contacts.

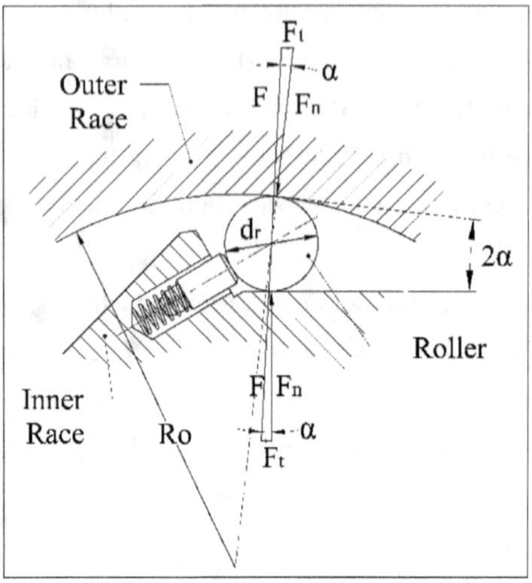

Figure 1.6. Free body diagram of clutch section for single clamping zone.

The contact stress is considered as one of the significant design parameters for OWC design and life cycle calculations. During freewheel conditions, the clutch unit becomes free in the reversed direction by executing relative motions between the races. **Figure 1.6** shows the free-body diagram of roller OWC section with a single clamping zone and the corresponding forces that develop during the engaged condition of torque transmission.

The force analysis indicates that the tangential force determines the transmissible torque. The transmissible torque (T) is calculated using **Eq. (1).**

$$T = z \times R_o \times F_n \tan\alpha \tag{1}$$

Where F_n is the normal force at the contact region, z is the number of clamping zones, R_o is the inner radius of the outer race, and α is the clamping angle. From the wedge friction theory applicable to the clamping zone, the following condition should be satisfied for positive torque transmission without any slippage **[9]**,

$$\tan\alpha < \mu \tag{2}$$

The friction coefficient (μ) varies in a specific range depending on the surface and lubricating conditions. Its value ranges from 0.08 to 0.12 for the hardened steel with general transmission fluid. Considering the conservative value of 0.08, the minimum clamp angle required from **Eq. (2)** is 4.574° **[10]**. Thus, the functioning of the OWC is dependent on the geometrical as well as the frictional conditions at the clamping zone. By operating the clutch repeatedly, the contact conditions progressively change due to wear, and the clutch eventually fails in fatigue.

1.1.3 Geometrical and material specifications

To achieve uniform loading and unloading of roller OWC, a smooth transition from engagement to disengagement of rollers must be attained simultaneously at all the contact regions. In order to transmit the torque, rollers must wedge tightly between the races in the engaged mode of operation, and such angle must not introduce any slip. A line drawn between the contact points of rollers with inner and outer races is called a strut, and the angle between this strut and the line joining the centre of the rollers normal to contact surfaces is called a strut angle 'α' as depicted in **Figure 1.6**. The strut angle is a critical parameter which defines the normal load at the contact points. There is a practical limit on minimum strut angle as the release of elastic strain energy tends to trap rollers and prevent smooth disengagement, and the applied torque imposes higher normal load and high stresses for small angles. Conversely, the larger the strut angles, the smaller the contact stresses, and the clutch becomes more sensitive to surface finish. Typical value of strut angle ranges from 3 to 5 degrees **[9-10]**.

The geometrical parameters **[6]** listed in **Table 1.1** for roller OWC specimens of different sizes are considered for the analysis. Three specimens are selected and classified based on their increasing torque transmission capacity, i.e. small, medium and large as specimens A, B and C, respectively. All three specimens are designed for the same clamp angle and similar surface conditions that play an essential role in achieving simultaneous engagement-disengagement, which results in uniform load distribution while transmitting the torque. The force analysis at contact regions is carried out using conventional Hertz cylindrical contact theory **[11]**, and the results are listed in **Table 1.2.**

Table 1.1. Geometrical parameters of OWC specimens [6]

Parameter	Specimen A	Specimen B	Specimen C
No. of rollers, z	8	8	10
Roller diameter (d_r), mm	6	7	12
Roller length (l), mm	7	9	12
Outer race ID (Do), mm	55	82	128
Inner race mean diameter (D), mm	42.7	67.6	103.5
Minimum clamp angle (α_{min})	3.2°	3.2°	3.2°

Table 1.2. Force analysis of OWC specimens

Parameter	Specimen A	Specimen B	Specimen C
Tangential load at contact (F_t), N	410	701	1661
Normal load at contact (F_n), N	7317	12542	29708
Resultant force (F), N	7328	12562	29755
Dynamic capacity (C), N	3470	5074	12404
Allowable contact stress [S_c], GPa	4.5	4.5	4.5
Maximum contact stress (S_c), GPa	3.7	3.9	4
Reduction factor (λ)	0.61	0.61	0.61
Material constant (M), $N/mm^{50/27}$	590	590	590

Since the contact regions of OWC are subjected to high contact pressure, the materials of rollers and races require proper treatments to improve mechanical properties. For this purpose, the races are manufactured from case-carburized steel (DIN 20MnCr5) and the rollers are made from hardened bearing steel (AISI 52100) to impose fatigue and wear resistance properties. **Table 1.3** and **Table 1.4** describe the required standard chemical compositions for the clutch materials and the corresponding measured values using an optical emission spectrometer for races and rollers, respectively. The case hardening depth of inner-outer races is obtained by case-carburization surface treatment, which increases the carbon content at the contact surface from ~0.2 to ~0.8% weights up to 1.5 mm of case depth.

Table 1.3. Chemical composition (weight %) of inner-outer races.

Chemical composition	C	Cr	Si	Mn	P	S	Ni	Mo
DIN 20MnCr5	0.17-0.22	1.00-1.30	≤0.40	1.10-1.40	≤0.035	≤0.035	0.08	0.01
Measured	0.20	1.032	0.30	1.21	0.029	0.017	0.06	0.01

Table 1.4. Chemical composition (weight %) of rollers

Chemical composition	C	Cr	Si	Mn	P	S
AISI 52100 Bearing steel	0.90-1.05	1.00-1.60	0.10-0.35	0.30-0.75	≤0.050	≤0.050
Measured	1.00	1.031	0.35	0.45	0.026	0.037

1.1.4 Contact fatigue and wear

Failure due to fatigue is generally observed in machine elements subjected to cyclic or repeated loading through progressive damage. Fatigue damage occurs in the areas of high-stress concentration due to surface imperfections, material inclusions and other microstructural defects within the material volume. Investigating the fatigue performance of OWC becomes significantly essential for reliable and safe functioning. The service life of OWC is not only dependent on the nominal loads based on induced contact pressures but also on the operating conditions. The operating conditions that have a significant impact on service life include the amount of load, the nature of the load, the frequency of loading, the speed at which the load actuates, and the conditions under which the transition of freewheel occurs.

The wedge elements of OWC only come into contact during the change in relative motion between contacting surfaces. When the inner race speed exceeds the speed of the outer race or vice versa, the clutch slowly loses contact, and there is no friction between them to be stated as a non-contact zone. As shown in **Figure 1.7**, there are two short time spans only when the wedging elements come into contact with the races. One is during acceleration at the starting period, and the second is deceleration at the end region of the graph. These are contact zones when the clutch only engages and mostly overruns during the rest of the time.

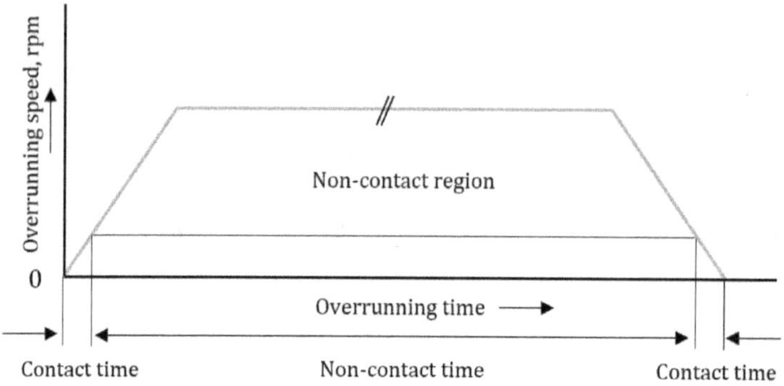

Figure 1.7. Graphical representation of the functionality of clutch w.r.t. time.

Therefore, when estimating the expected life of the clutch, it is necessary to consider two service conditions with respect to actual application, which are as

1. Fatigue life for an engaged condition
2. Wear life for an overrunning condition

When the clutch engages, the contact stress is induced at the contact surfaces, as shown in **Figure 1.8,** which is directly proportional to the engagement torque. The fatigue caused by such repeated engagement will lead to surface pitting of the contact surfaces. Also, when the applied load changes during the single engagement or variable/vibrational loads encountered during the engagement, the effect of increased overall load values must be considered while determining the service life of the clutch. In order to determine the number of load cycles, the Wohler curves (S-N curve- service life curves) are utilized. Due to the frequent loading-unloading function in OWC applications, the cyclic contact stresses are developed in the vicinity of the contact geometries, which causes the clutch to eventually fail after a number of load reversals. It is, therefore, necessary to perform endurance tests to evaluate the fatigue life of OWC.

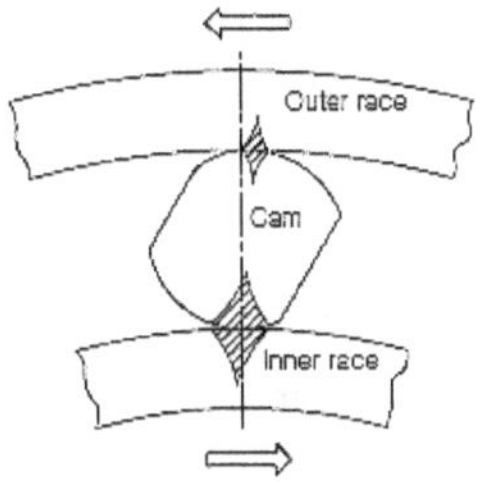

Figure 1.8 Contact condition during engagement

Similarly, on the other side, OWC works as an overrunning tribological device during most of its usual operation, where the wedging elements are disengaged from the mating elements and rotate with relative velocities. Wear of the contact geometries due to long overrunning periods is one of the principal failure modes of this clutch. After a certain period, the contact surfaces are worn out due to continuous overrunning operation; therefore, it is of prime importance to estimate the overrunning wear life as well.

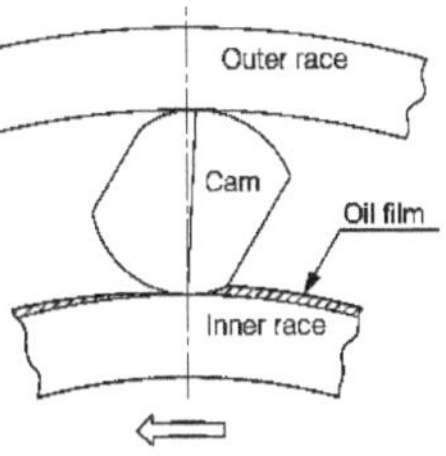

Figure 1.9 Contact condition during overrunning

When the clutch freewheels, the contact surfaces, as shown in **Figure 1.9,** skid with respect to each other, which is directly proportional to the overrunning speed. Therefore, it is necessary to pay attention to any foreign particles or debris present in the vicinity of the contact surfaces, which may cause surface damage due to wear. Since the contact pressure due to weak spring force is low and due to the presence of sufficient lubrication, the wear of the parts can be delayed, but the wear life of OWC must be estimated, especially for applications involving high overrunning speeds for longer durations.

1.2 Life Estimation of OWC and their Measurements

The fatigue and wear life estimation of OWC under specific loading and operating conditions should be carried out using experimental and numerical techniques. The estimated experimental fatigue and wear lives are required to be validated with the numerical lives obtained through numerical analysis. A life analysis approach and its measurement techniques are discussed in the following sub-sections.

1.2.1 Fatigue life estimation

The cyclic fatigue experiments on OWC specimens are required to be performed on customized test rig. Such test rigs must be designed and developed to exhibit the required loading-unloading operations on clutch specimens. The life cycle counts after each failure can be recorded using a PLC-based measuring system. The endurance tests should be performed to acquire the life data on this test rig. The experimental fatigue life shall be evaluated using a 2-parameter Weibull estimate based on the inverse power model. Later, an FE-based numerical model has required to be prepared using a software package to study the fatigue behaviour under similar operating and loading conditions to estimate the numerical fatigue life. The numerical model utilizes the concept of damage mechanics to evaluate the fatigue life based on crack initiation and propagation under cyclic loading.

1.2.2 Wear life estimation

On the other side, a pin-on-disc test rig for wear experiments can be used to evaluate the experimental wear life of the clutch. A specific clamping device must be developed to replace the standard rig contact elements thereby exhibiting the similar contact conditions of OWC contact regions. The experimental wear analysis should be carried out under specific loading and operating conditions of the clutch to estimate the wear rate. Thus, the experimental wear life shall be estimated using the wear rates. An FE-based numerical analysis has required to be performed using a software package to estimate the wear rates numerically under similar operating and loading conditions using Archard's wear model. Thus, the numerical wear life shall be estimated using the wear rates and validated with experimental results.

1.3 Indian Knowledge System

The fatigue and wear life investigations of OWCs through experimental methods are essential to examine the behaviour of OWC under various operating conditions. The power transmission and automobile industires **[6]** across India have improvised their test capacities to conduct extensive fatigue tests for automobile components including OWCs in order to validate their product compatibilities, durabilities and product optimization. There have been an extensive array of Indian research articles, describing contact fatigue mechanisms observed for various applications involving bearings, gears and OWCs. Moreover, these studies have also introduced diversed methodologies to perform numerical analysis on fatigue and wear problems. Contributions made by such Indian researchers are cited in Chapter 2 and other relevant sections of the thesis.

1.4 Motivation

While performing its function, the one-way clutch is always subject to a high amount of repeated contact stresses and hence is under the influence of high cycle contact fatigue, ultimately leading to failure. In order to better interpret component failures and accurately predict fatigue life under harsh operating conditions, a detailed understanding of fatigue damage is required. This proposal offers a step forward in understanding fatigue damage behaviour for products like OWCs because of the complex geometry of contact regions. Many researchers have studied contact fatigue in the past. Also, significant knowledge is present on this subject, particularly regarding the bearings. Despite this, very few efforts have been made to investigate the same for OWC applications. Investigating the fatigue and wear life of OWCs is crucial, particularly under diverse and unpredictable loading conditions, which are encountered more often in various industrial applications of OWCs. Many existing research studies rely on empirical life models that may not fully capture the complex mechanisms involved in fatigue and wear. These models often require extensive experimental data and may lack generalizability across different clutch designs and applications. Also, the previous studies are not an outcome of the experimental study capturing the real-life loading scenario. It makes it difficult to compare the fatigue life results from different studies and apply findings universally. Investigation is required to be carried out to understand the fatigue and wear behaviour of OWCs. In addition to this, the surface interaction between specific materials for OWCs under various operating conditions may not be fully explored. Understanding the wear mechanism is vital for predicting wear life accurately. Therefore, the present work is focused on providing a comprehensive study of the fatigue and wear behaviour of OWC under various operating conditions. This continued research is also essential for ensuring the reliability of the design of the OWCs.

1.5 Scope of the Dissertation

The dissertation mainly focuses on estimating and validating the fatigue and wear life of mechanical one-way clutches, for which the following activities have been defined and executed as a part of the research work.

- The fatigue life estimation of OWC using experimental techniques.
- The study of fatigue damage mechanisms for OWC.
- Validation of experimental fatigue lives using a finite element-based numerical model.
- The experimental investigation of wear behaviour and estimating wear life of OWC.
- Validation of experimental wear lives using a finite element-based numerical model.

1.6 Organization of Thesis

The thesis introduces the development of an experimental test rig for performing fatigue experiments on one-way power transmission products. The conventional fatigue life model is applied to estimate the load-life relationship for OWC. The accelerated tests are performed to collect the life data, and the load-life relationship is developed using a 2-parameter Weibull estimate using the inverse power law model.

The experimental load-life model is then compared with the numerical model, where the damage-coupled FE model is employed to predict the total life. The wear life of OWC is also predicted by performing wear experiments on conventional pin-on-disc tribometer. The experimental wear results are then validated using the FE-based Euler's integration scheme based on Archard's wear model.

CHAPTER 1 Introduction

This chapter introduces one-way power transmission elements, the theory of operation, contact stress estimation, and the need to evaluate fatigue and wear life.

CHAPTER 2 Literature Review

This chapter highlights the recent literature on contact fatigue, accelerated fatigue life models, fatigue and wear life prediction of rolling contact elements and their validation using finite element-based numerical models. The summary of the literature emphasizes the importance of evaluating fatigue life, especially for the application of high cycle fatigue.

CHAPTER 3 Design And Development of Experimental Test Rig for Fatigue Life Estimation of OWC

This chapter introduces an experimental test rig designed and developed for performing accelerated life tests on OWC specimens. The test rig specifications, design of test components and test rig validation are also discussed.

CHAPTER 4 Experimental Test Plan and Test Methodology for Fatigue Life Estimation of OWC

This chapter demonstrates the test plan and test methodologies for executing accelerated life test analysis (ALTA) for estimating fatigue lives of OWC using experimentally collected life data. The load-life relationship is developed using a 2-parameter Weibull estimate based on the inverse power model. The conventional Lundberg-Palmgren fatigue life model is applied to OWC specimens for different sizes, and the life results are compared with the Weibull model. The effect of contact angle profiles and contact pressure on fatigue lives are also investigated. Finally, the failure analysis of the OWC specimen is carried out using a scanning electron microscope (SEM).

CHAPTER 5 Numerical Analysis for Fatigue Life Estimation using Continuum

Damage Mechanics based Finite Element Model

This chapter explores the concept of continuum damage mechanics to study the crack initiation and propagation life under high cycle loading. An FE-based material model for one-way line contact is prepared for the contact analysis. A numerical strategy is discussed using contact analysis results to evaluate the crack initiation and propagation life. The life results using a numerical model are compared with experimental results. The effect of contact pressure, case carburization and residual stress on fatigue life is investigated.

CHAPTER 6 Experimental Wear Life Estimation of OWC using Pin-On-Disc Tribometer and their Validation using Finite Element Model.

This chapter includes the experimental approach to perform wear experiments on OWC contact geometries using a pin-on-disc tribometer. The experimental wear results for specific loads are utilized to estimate the wear amount at operating load levels of the clutch. Apart from that, the coefficients derived from the experiments are used to perform finite element simulation to predict the wear based on Archard's wear model. The wear life is then predicted for various relative speeds of OWC.

Conclusion and Scope for Future Work

This chapter presents the conclusions derived from the experimental and numerical studies for evaluating fatigue and wear lives of OWC. The scope for future work is also discussed.

Chapter 2

Literature review

This chapter highlights essential literature on OWC applications, fatigue and wear life models, experimental and numerical investigations related to contact fatigue and wear measurements. Research gaps are identified based on the literature review. The research objectives for the current study are framed and followed by the research methodology to carry out the research work.

In the mechanical design of any component, fatigue damage has been associated with the magnitude of stress level and the number of stress cycles undergone by the component or the system under investigation. In the past few years, many investigations have attempted to describe the physical mechanisms implicated by contact fatigue and proposed different models to predict the fatigue lives of bearings. Based on a similar approach, the current chapter pitches some insights on important literature works published regarding OWC: design and analysis, contact fatigue, wear, experimental and numerical test methodologies and other relevant studies for estimating fatigue and wear life of OWC.

2.1. One-way clutch: Historical Background

OWCs were used in passenger car transmission systems in the early 1960s. In 1960, Sauzedde **[12]** presented a paper that was considered the benchmark technical document on roller OWC analysis and design. The equations for contact stress, outer race circumferential stress, and roller centrifugal force computations were derived from the work. The further development in the design of OWC from its inception made their usefulness in other industrial segments **[13-14]**. The contact mechanics of OWC was critically described by Haka et al. **[15]**. Several improvements refining the initial equations **[16]** have been made and published pursuant to this seminal paper. Kremer **[17]** and Chesney et al. **[9-10]** have defined the static and dynamic behaviours of OWC, the practical applications and proposed geometrical operational limits, which are considered the current state-of-the-art for designing OWC. Brief design equations and analysis of OWC, including the theory of operation, equilibrium equations, system deflections and stress equations for the design of roller OWC, were described by

Fanella [18]. Later, a standard design formulation for OWC was established and defined to be used as a general purpose [19]. Wei Xue et al. [20] have derived an optimization model based on the mechanism of a one-way roller clutch to achieve maximum torque capacity and minimum system stresses by introducing a constant strut angle, which is considered as the significant design parameter of OWC. The effect of contact parameters on OWC operation is further studied and explained by Vernay et al. [21]. The recent development and optimization of OWC made its successful implementation in starter drives and power transmission industries [22-25], where OWCs are operated under heavy loads. Future trends and advancements in smart manufacturing have led to the evolution of OWC for automotive industries [26]. In order to guarantee the reliability of such products, it is important to evaluate the fatigue life of OWC through endurance tests at various operating conditions.

2.2. Life Prediction Models for Contact Fatigue

Generally, contact fatigue observed in OWC applications differs from classical fatigue, where two non-conformal contact geometries define the contact region, and the state of stress is governed by Hertz's contact stress theory [11]. Material fatigue is a dominant failure mode for OWC if adequately assembled, loaded and lubricated. The highly localized contact regions of such non-conformal line contacts are the main reason for introducing fatigue wear due to the partial sliding motion in OWC races. Due to repeated engagements under load, the highly stressed contact regions undergo smearing of contact surfaces, leading to surface damage in terms of spalls. Over the years, several empirical and research models have been developed to predict contact fatigue lives for roller element bearings (REBs), and an extensive review has been carried out by Sadeghi et al. [27]. All such models are developed and differ from one another in their distinct modelling considerations and various aspects of contact fatigue in REBs. The initial efforts were made to derive the formulas for computing fatigue lives from full-scale or endurance tests [28-29], generally identified as empirical models. These models neither consider the constitutive behaviour of materials under contact nor the plastic strain and residual stress effects. On the other side, the research models require stress-strain data and are prominently theoretical in nature. In recent years, finite element-based computational models have been utilized in critical fatigue life investigation where material microstructure is explicitly required to be considered. A detailed description of such life models is provided in the following subsections.

2.2.1. Conventional life models

The conventional life models are also known as classical/empirical or engineering life models. The first theoretical basis for formulating the contact fatigue life was reported by the seminal work of Lundberg and Palmgren **[30-31]** to predict bearing lives. The Lundberg-Palmgren (L-P) theory evaluates the survival probabilities of bearings subjected to repeated loads, which is the function of maximum orthogonal shear stress, the depth at which it induces, the stressed volume as in **Figure 2.1** and the number of fatigue cycles. Such a bearing life equation was formulated using the Weibull theory **[32-33]** of probability to aid in the reliability aspects. It is considered as a significant step forward in the accomplishment of predictive methods for bearings and similar sort of applications.

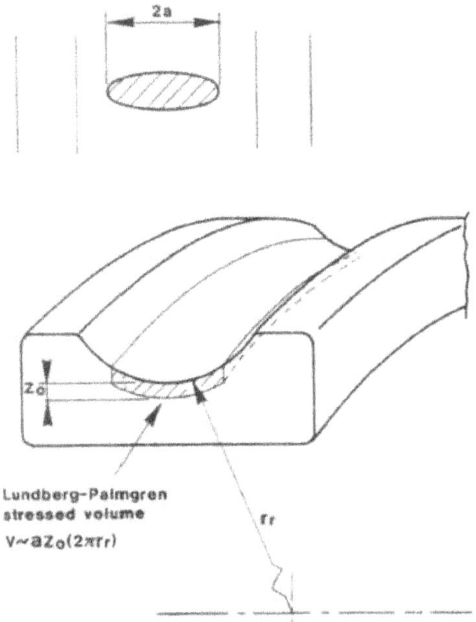

Figure 2.1 Lundberg-Palmgren stressed volume **[30-31]**

Chiu et al. **[34]** used the Weibull probability function to observe the fatigue life scatter experimentally; however, the underlying uncertainty in identifying the physical mechanisms responsible for contact fatigue could not be explained, which led to the formation of research models.

Despite the wide recognition of L-P theory, it has a few limitations, such as overlooking the presence of surface lubrication and surface-initiated failures. The loading at the contact regions was assumed to be pure normal without considering surface shear traction. The contact surface was assumed to be perfectly smooth. However, practical lubrication is provided occasionally. Also, there is some evidence of surface-initiated failure; some surface shear traction is always present, translating the maximum orthogonal shear stress closer to the surface and surface irregularities (scratches) differ the stress field from the pure Hertzian stress field. Ioannides et al. **[35]** proposed a model which has become a well-accepted basis for estimating fatigue lives as a further development in bearing life prediction to overcome the limitations associated with the L-P model with some modifications. The first innovation was the consideration of stressed volume at the sub-surface as an array of relatively small volume elements, each with its probability of survival experiencing individual local stresses. Secondly, they identified a threshold value of stress, a fatigue limit for structural components below which no failure was expected to occur. In such a manner, local stresses and other effects stemming from surface stress concentration due to surface roughness, dents, and edge stresses were considered in life predictions.

Schlicht et al. **[36]** proposed a bearing life model based on the principles that the critical stress leading to the failure is the von Mises stress, and the governing factors that influence the life of bearings are residual stresses and plastic flow arising during contact fatigue. Tallian **[37-38]** presented a fatigue life statistical model for rolling element bearings. This model utilized the orthogonal shear stress field to compute fatigue lives as the crack propagation life according to Paris law, neglecting crack initiation life. It should be emphasized that the L-P equation describes the critical stress-life exponent as a Weibull slop for a given probability of survival, and this exponent depends upon dispersion in life data. Based on that, Zaretsky **[39]** proposed a similar model to L-P with modifications. The dependence of the stress-life equation on Weibull slop was eliminated, and dependence on the depth term was also eliminated. The critical stress value utilized was not the orthogonal shear stress but the maximum shear stress.

Later, Harris et al. **[40]** compared bearing lives using L-P and Ioannides-Harris theories, where the ratios of predicted lives and experimentally observed actual lives were compared. Otsuka et al. **[41]** examined the effect of considering surface friction force in life prediction that causes the maximum shear stress depth to move closer to the contact surface. Further, Kudis et al. **[42]** proposed a statistical life model including the effects of parameters like contact size, contact pressure, coefficient of friction, etc., on fatigue lives. In addition, Harris et al. **[43]** further introduced the Ioannides and Harris method more generically. Their theory, which was still based on the L-P theory, employed a stress-life factor to modify the fatigue life predicted by L-P. This stress-life factor incorporates the effect of applied or induced residual stresses with empirically determined fatigue stresses. The loss of hardness at higher temperatures was also incorporated into the bearing life equation. A schematic timeline representing the author's contribution in the context of the engineering life model is illustrated in **Figure 2.2**

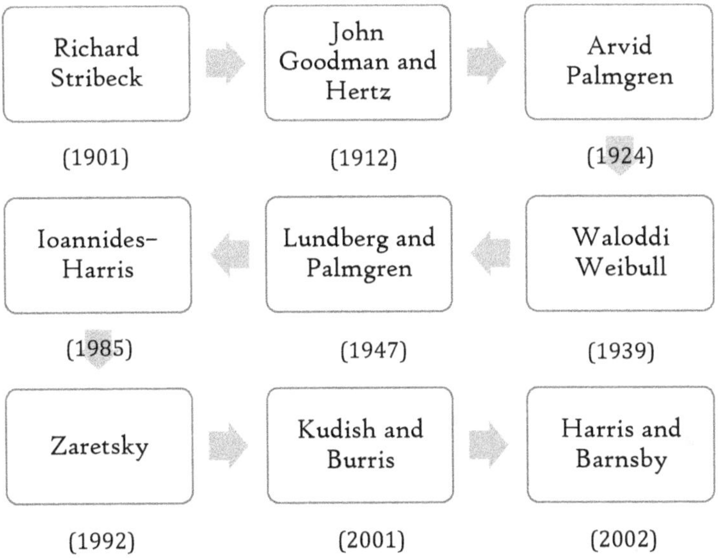

Figure 2.2 The author's contribution representing the research in the context of engineering life models

In view of all these engineering models, the fatigue life estimation methodologies of bearings were investigated in detail by various authors and readily available in the form of ISO standards **[44-47]** and ANSI-based load-life equations **[48-49]**, where the multiplicative factors were introduced for rolling bearing life to account for different levels of material fatigue properties, reliability and lubrication. ISO 281:2007 is the internationally recognized standard for predicting the life of bearings using L-P theory. The same is applied to estimate the life of roller OWC.

2.2.2. Research life models

Apart from conventional fatigue life models to estimate fatigue life, the research models have been proposed over times where the surface and sub-surface originated spalls were manifested as governing failure mechanisms for fatigue damage **[50]**, as depicted in **Figure 2.3.** Both these failure mechanisms are described by micro-cracks initiated from micro-voids present in the material volume, propagated to the surface through the critically stressed material regions.

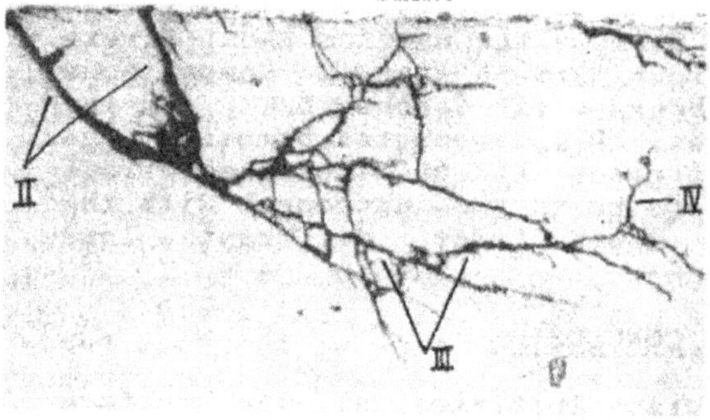

Figure 2.3 Surface and sub-surface originated cracks under contact fatigue **[50]**

Generally, the research models utilize stress-strain data of the materials in contact. Such models are bound to a distinct form of failure: crack initiation-propagation or surface damage. Keer et al. **[51]** and Alley et al. **[52]** have attempted to characterize crack initiation and propagation stages without accounting for the stochastic nature of contact fatigue. A new advanced modelling approach was developed recently to overcome this limitation by incorporating microstructure topology using the Voronoi tessellation technique. The scatter in fatigue lives can be taken care of using this technique by incorporating microstructure randomness in the material domain. All these approaches can be used to predict fatigue lives. However, fatigue damage evolution cannot be described without coupling the contact stress field with fatigue damage. A concept of continuum damage mechanics (CDM) was proposed by Lemaitre **[53]**, which utilizes a framework to capture the crack initiation and propagation lives along with damage evolution for fatigue life prediction. Damage mechanics coupled fatigue life models were later successfully adopted to predict various fatigue problems. Raje et al. **[54]** and Jalalahmadi et al. **[55]** applied this concept to a 2-dimensional framework of line contact to estimate fatigue lives and found a good correlation of estimated scatter/spalling patterns with experimental results. The uniaxial fatigue tests **[56-57]** of tensile specimens were conducted using this concept to predict fatigue lives.

Consequently, using this approach, 3-dimensional Finite Element (FE) models were proposed [58-59] for predicting fatigue lives under point and line contacts. Warhadpande et al. **[60]** and Bomidi et al. **[61]** have incorporated the effect of plastic strains to obtain stress histories accurately under a large number of contact loads. Additionally, the elastic-plastic behaviour of the material under non-linear kinematic hardening effects was studied **[62-63]** to solve the fatigue problems. Slack et al. **[64]** and Walvekar et al. **[65]** have implemented this concept to study fretting fatigue. Li et al. **[66]** modified a damage mechanics model for measuring accurate contact stress variations due to repeated contact loads. Ahmadi et al. **[67]** have performed a numerical investigation to predict torsional fatigue lives for nickel-based alloys using CDM. Using CDM, Golmohammadi et al. [68] utilized a coupled multi-body elastic-plastic FE model to investigate surface defects on rolling contact fatigue. Vijay et al. **[69]** simulated crack initiation and propagation stages for polycrystalline materials to estimate the fatigue life of bearings using a damage mechanics framework. Park et al. **[70]** further proposed a microstructure-based crystal plasticity model in conjunction with CDM to validate bearing steels' rolling contact fatigue life. In order to improve fatigue lives in contact fatigue, surface hardening is one of the commonly employed techniques. Scott **[71]**, Sharma **[72]**, Widmark et al. **[73]** and Bhattacharya et al. **[74]** have analysed experimentally the effects of hardness on fatigue lives, but very few were analytically verified. Walvekar et al. **[75]** and Golmohammadi et al. **[76]** investigated the effects of case-carburization on fatigue life using the damage mechanics model. Thus, research models provide a more comprehensive and realistic approach to investigate fatigue damage and predict fatigue life more accurately based on the available governing failure mechanism.

2.3. Failure Mode in Contact Fatigue

The process of contact fatigue was divided into three stages: incubation of cracks, followed by stable propagation and branching out period of the cracks, as displayed in **Figure 2.4.** Shao et al. **[77]** used dislocation theory to explain crack initiation at the location where the maximum value shear stress occurs.

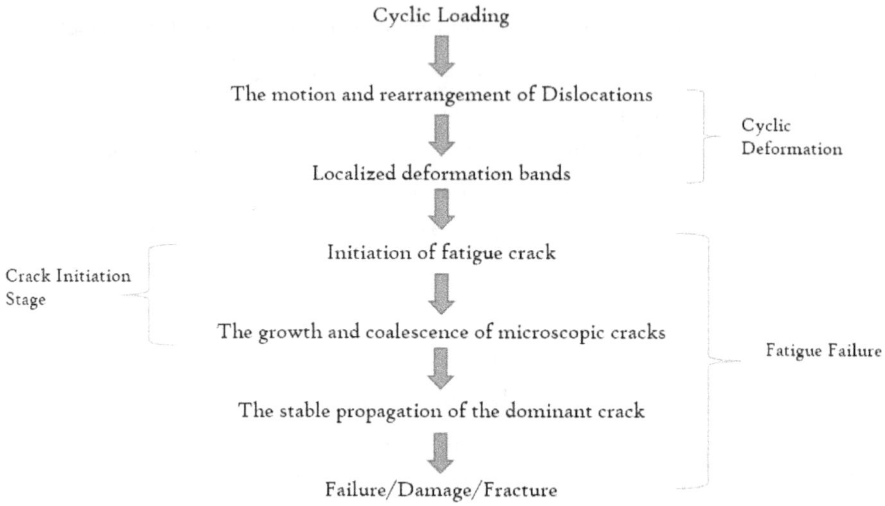

Figure 2.4 Failure process under contact fatigue **[50]**

The non-conformal contact between OWC elements makes contact fatigue different from classical fatigue due to the complex nature of multi-axial stress induced under the contact surface. The principal stress direction at a material point is not fixed under the load cycle. Such localized nature of contact increases the effect of heterogeneous micro-scales on the contact mechanism, leading to scattering in fatigue life estimation **[78].** Another important aspect of contact fatigue in the case-carburized OWC is the micro-plastic deformation at the micro-scale level. Therefore, fatigue in OWC due to repeated loading and unloading, in general, causes the irreversible plastic flow of the material at the micro-scale and consequently results in damage accumulation. Under this action, the fatigue cannot be witnessed until these micro-cracks propagate, resulting in macroscopic failure at the surface. The subsurface shear component that alternates under cyclical load describes the fatigue strength of the materials in contact, as reported by the authors **[79-80].** So, the shear stress range and the accumulated plastic strain are considered the governing factors for life prediction. Thus, subsurface-initiated damage, in addition to surface damage due to smearing under heavy loads, stands as the leading cause for failure and limits the life of OWC due to changes in geometry responsible for torque transmission.

2.4. Accelerated Life Test Analysis (ALTA)

There have been very few studies that empirically validate the fatigue life of OWC by performing endurance tests. The life cycle analysis for OWC was not performed until Durkopp et al. **[81]** executed tribological experiments in the laboratory to study tribological aspects of roller OWC fatigue for an individual clamp zone under analysis. Later, Zhifang et al. **[82]** derived the dynamic wedging characteristics of OWC, which is a crucial literature for performing load tests. Conventional fatigue life analysis involves analysing life data under standard operating conditions to estimate the life characteristics of a product or system under investigation. For most of the reason, it becomes difficult to obtain these data due to the long lifetimes of the products. Therefore, it is necessary to force the product to fail faster than it would under standard operating conditions to analyse the relevant failure mode and estimate its life characteristics. In other words, the failure modes need to be accelerated, and hence, the analysis has been specified as Accelerated Life Test Analysis (ALTA).

Since the fatigue life of OWC is described by high cycle fatigue (HCF) **[83]**, it would be more feasible to accelerate the load levels using ALTA to get quicker failures. Accelerated life tests are performed to get information about the failure time distribution of products such as HCF. The accelerated test methodologies were first formulated by Nelson et al. **[84]** to estimate the reliability of any product from experimentally available data and predict its life at operating conditions. Further advancement in accelerated testing was reviewed by Meeker et al. **[85-88]**. The reliability prediction under accelerated testing was further extended using various statistical models **[89-91]**. The accelerating model selection procedure and parametric estimation method were presented along with test guidelines **[92-94]**, as displayed in **Figure 2.5.** Pranav et al. **[95]** have presented the guidelines for performing accelerating tests for various mechanical components in controlled laboratory conditions. Nam et al. **[96]** performed an accelerated analysis using temperature as an accelerating variable to estimate the life-temperature relationship for bearing lubricants and evaluate the effect of temperature on bearing lives. Later, an optimum test plan design using different accelerated test strategies was proposed by Chen et al. **[97]**. Cai et al. **[98]** have studied the effect of increasing load frequency and stress amplitude on the accelerated fatigue tests, where they have concluded that the increase in load frequency may change the material conditions due to a rise in temperature owing to the conversion of plastic work into heat. Therefore, it is concluded that ALTA is an appropriate and well-proven method to estimate the fatigue life from experimentally collected data under HCF.

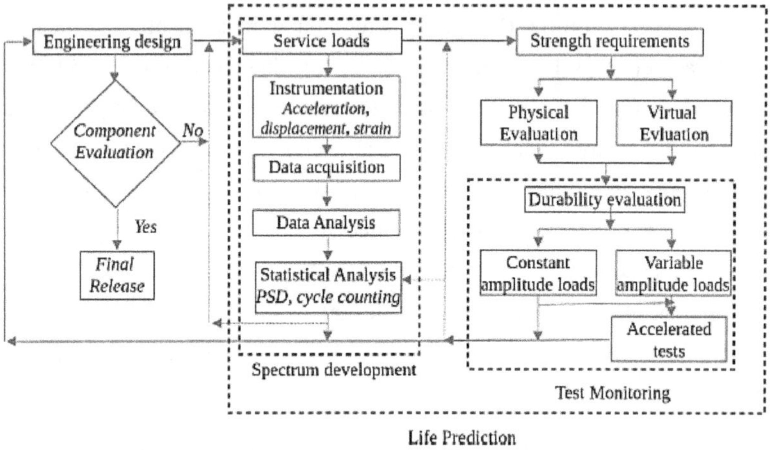

Figure 2.5 Life prediction using ALTA methodology **[92]**

2.5. Surface Analysis and Wear Measurements for Contact Bodies

Apart from the fatigue life for engaged conditions, wear of the contact geometries due to long overrunning periods is also one of the principal failure modes of the clutch. Wear in overrunning clutch depends on the pressure and the sliding velocity of contacting surfaces. Both these parameters considerably vary as per the relative speeds between the inner and outer races of the clutch. Wear can be evaluated either in terms of mass loss or change in the geometry of the contact bodies **[99-100]**. In the early stage of development, Dürkopp et al. **[101]** examined the friction and wear behaviour of one-way freewheels by experimental testing. Many authors have explored the contact characteristics of one-way clutches over the years. As a part of this, the frictional characteristics of a one-way clutch under transient load and the analysis showing the effect of fluctuation of input speed on wear life during the transmission state were analysed by Li et al. **[102]**.

Wear is a material removal process by surface degradation due to sliding or rolling motion exhibited by two contact surfaces. The performance and reliability of any mechanical component are characterized by the friction losses displayed through various phenomena like adhesion, abrasion, wear, fatigue, fracture, etc. Several models based on such driving mechanisms were developed in literature **[103-107]**. A linear relationship between normal force and frictional force is defined by a proportionality constant, i.e. coefficient of friction as per Coulomb's model, which is applicable for low contact pressures essentially observed in sliding contact pairs by Xincai **[108]**. Lim et al. **[109]** investigated that similar materials exhibit different wear mechanisms in different wear conditions. It is difficult to identify a generalized theory that can be single-handily applicable to all wear problems with every wear condition by Czichos et al. **[110]**. Archard's wear theory of sliding contact is one of the famous phenomenological wear models that estimates wear volume based on the normal force and sliding distance between two contacting bodies. This model requires parametric calibration from experimental results and is still the most reliable for studying the wear behaviour of tribological systems.

In order to investigate the tribological performance of interested materials in contact with sliding pairs, a pin-on-disc tribometer has been a widely used experimental facility. Apart from that, this kind of relatively simple test setup is employed to evaluate the wear rate of materials and to identify the favourable parametric conditions which improve the life of the components or even to assess the effects of various lubricants on the operating conditions of the mechanical components. Though the wear mechanism and contact parameters considered on pin-on-disc can differ from actual practice, an FE simulation strategy is utilized to compare the experimental wear results. In the last few years, many efforts have been made to develop FE simulation strategies for wear prediction. An FE iterative approach has recently been utilized to update the contact geometries with each wear step. The contact pressure and sliding distance are also altered accordingly. A wear model typically known as the global incremental wear model was developed by Hegadekatte et al. **[111-113]**. The wear depth for the pin-on-disc test using FE simulation was determined in accordance with Archard's wear model.

The FE post-processor was utilized to verify total wear for corresponding contact pressure values at predefined time intervals. Wear was calculated by a change in the element height of contact surfaces, and re-meshing of the worn-out geometry was applied using this approach. Martinez et al. **[114-115]**, Arjmandi et al. **[116]**, Maier et al. **[117]** and Zhang et al. **[118]** utilized the UMESHMOTION subroutine in conjunction with the adaptive mesh technique to simulate the material removal in sliding contact. Adhithya et al. **[119]** and Haoyuan suo et al. **[120]** utilized an adaptive mesh simulation technique to predict material removal profiles using Archard's model. The contact pressure distribution was also verified by the experiments. Tavoosi et al. **[121]** performed pin-on-block experiments for nanostructured NiAl coating to identify wear characteristics under various loads and compared them with numerical simulation. Podra et al. **[122]** and Hatam et al. **[123]** compared the wear results of pin-on-disc experiments with the FE model to verify the wear life of components using the linear wear model. The wear transition observed from mild to severe depends on normal load, sliding speed and hardness of contact surfaces. Yu et al., Stanković et al. and Jain et al. **[124-126]** utilized Archard's wear model to evaluate wear parameters for bearings. Markus and Andre **[127-128]** have analysed air foil thrust bearings to predict the wear of the coating layer using a similar approach. Ashraf et al. and Jeon et al. **[129-130]** suggested that the FE simulation results are sensitive to the wear rate value with respect to sliding distance. Xue et al. **[131]** performed a 3-dimensional FE wear analysis to predict the service life of self-lubricating spherical plain bearings, and the results were compared with the experimental data. Later, various authors **[132-139]** investigated the effects of preload, surface roughness, rotational speed, manufacturing methods and lubrication on the wear behaviour of bearings using a numerical model. Bhoi and Singh **[140]** performed dry sliding wear tests using weight loss techniques to estimate the wear rate for Al-Y2O3-based composite material.

2.6. Summary of Literature Review

It is evident from the available literature that the fatigue life of any mechanical components exhibiting line or point contacts can be estimated using conventional and research fatigue life models. The available fatigue life models have been utilized to study material failure and predict fatigue life under various loading and operating conditions of bearings, gears, etc. As such, OWC comes under the category of HCF, the extensive endurance tests have not been feasible, and an ALTA can be proven to be a more appropriate method to estimate fatigue life from experimental data. Fatigue damage can be studied using a specific test rig to execute the qualitative tests on OWC specimens. The conventional fatigue life model with ALTA can provide valuable insights into fatigue life prediction for available clutch geometries through experiments. However, an appropriate research model coupled with modern computational methods could predict and validate the fatigue life based on the failure mechanism observed through experimental tests. It is further known from the available literature that the pin-on-disc experimental test setup is the most widely used tribological device to study the wear and friction characteristics of sliding contact bodies. The same test setup can be utilized to estimate the wear life of OWCs experimentally. However, a specific clamping device must be designed to exhibit OWC contact conditions. Since the specific surface treatments are introduced to OWC materials, the corresponding friction and wear characteristics must be evaluated for the relevant loading and operating conditions to estimate the wear life of OWCs. In the last few years, an FE-based numerical model coupled with an adaptive mesh technique has emerged as the most prominent numerical strategy to evaluate the wear between sliding contacts, and the same can be utilized to perform a numerical investigation to estimate the wear life of OWCs.

2.7. Research Gaps

Based on the available literature that has been reviewed for the current study, it was found that great efforts have been put into the design, analysis and optimization of OWC by the researchers **[9-10, 12, 16-20]**. However, very few studies have empirically predicted and validated the fatigue life of OWC by performing endurance tests. There are many geometrical similarities between bearings and OWCs regarding manufacturing attributes, making them difficult to distinguish under direct observation. The fatigue life investigation for rolling element bearings involves performing fatigue experiments on a test rig comprised of a rotating shaft with pedestal bearing supports and a central load acting in a radial direction for load reversals. However, the same test rig cannot perform the fatigue experiments since OWC is functionally different from bearing regarding fatigue operations. Also, most of the previous investigations **[27-31]** have been carried out to analyse bearing failures that resulted from the rolling action of the balls/rollers, but not specifically for OWCs. Therefore, a test setup is specifically required to be designed for performing fatigue experiments under service load conditions. Simulating the service loading conditions by developing a customized test rig shall result in the realistic estimation of the fatigue life of OWC instead of considering a single clamp zone under investigation. Also, the earlier engineering models **[34-39]** have been studied for various applications, specifically for the bearings, but very few engineering fatigue life models were proposed to OWC and were not validated through extensive parametric investigation.

Moreover, experimentally observed bearing lives follow Weibull distribution closely **[30-33]**, but no probabilistic distribution has been applied to OWC for probabilistic life estimation. Similarly, several research works have been carried out in the past few decades to establish a generalized wear equation **[103-104, 111-113]**, which will be able to predict the wear rate on different tribological systems, but no wear model was found to be developed strictly for the long period overrunning application like OWC. Additionally, most of the early research was carried out to understand the wear characteristics of the clutch during load transfer when sprags are in contact, but no such study has been carried out for the overrunning function of the clutch.

2.8. Research Objectives

From the literature review, it was observed that the endurance tests for OWC have not been performed so far for the real-life operating and loading conditions on a specific test rig. Several works have been done to optimize the design of OWC based on the theoretical aspects, but no specific experimental investigations have been carried out to study the fatigue and wear behaviour of OWC. Also, various numerical models have been available for contact fatigue and sliding wear applications that can be explicitly applied and studied for fatigue and wear life estimation. However, no efforts have been made to employ such models specifically for OWC applications.

In light of the above discussions, the following objectives are framed as guidelines to execute the current research work:

1. To design and develop an experimental test rig to perform fatigue experiments to estimate the fatigue life of one-way power transmission elements.
2. To estimate the fatigue life of OWC using experimental life data considering the effect of various parameters on fatigue life, such as load and cam geometry.
3. To investigate the governing failure mechanism for fatigue damage resulting

from cyclic load on OWC.

4. To employ a numerical model for estimating the fatigue life of OWC to better characterize the surface–induced damage from the subsurface contact fatigue for crack initiation and propagation lives using the continuum damage mechanics approach and their validation using finite element analysis.
5. Estimating the wear life of OWC by performing wear experiments considering the effect of various parameters on wear life, such as load and sliding velocity, and their validation using finite element analysis.

2.9. Research Methodology

The present work mainly focuses on estimating the fatigue and wear life of OWC by experimentally investigating the fatigue and wear behaviour under various loads and operating conditions on specifically developed test rigs. The experimental results are validated with numerical results for applied numerical models by performing finite element analysis. As shown in **Figure 2.6**, a research methodology has been followed to execute the present research work.

The fatigue life models are applied to OWC applications where endurance tests must be performed on a functionally unique test rig. Therefore, a test setup is specifically designed and developed to induce directional dependent load reversals as per the one-way functionality of the clutch. The selection and design of the test rig components are required to be carried out using applicable machine design principles, followed by the fabrication and final assembly of the test rig components. The experimental fatigue life data are acquired from this test rig using endurance tests. The accelerated life test analysis is a well-recognised technique for performing high-cycle fatigue tests on the available test rig. The fatigue life for OWC is estimated from the experimental fatigue data for an appropriate load-life relationship using a 2-parameter Weibull estimate. The effect of contact pressure, contact angle and surface conditions on fatigue life are required to be estimated. Studying the governing failure mechanisms for fatigue damage caused under repeated loads from the endurance tests is also essential.

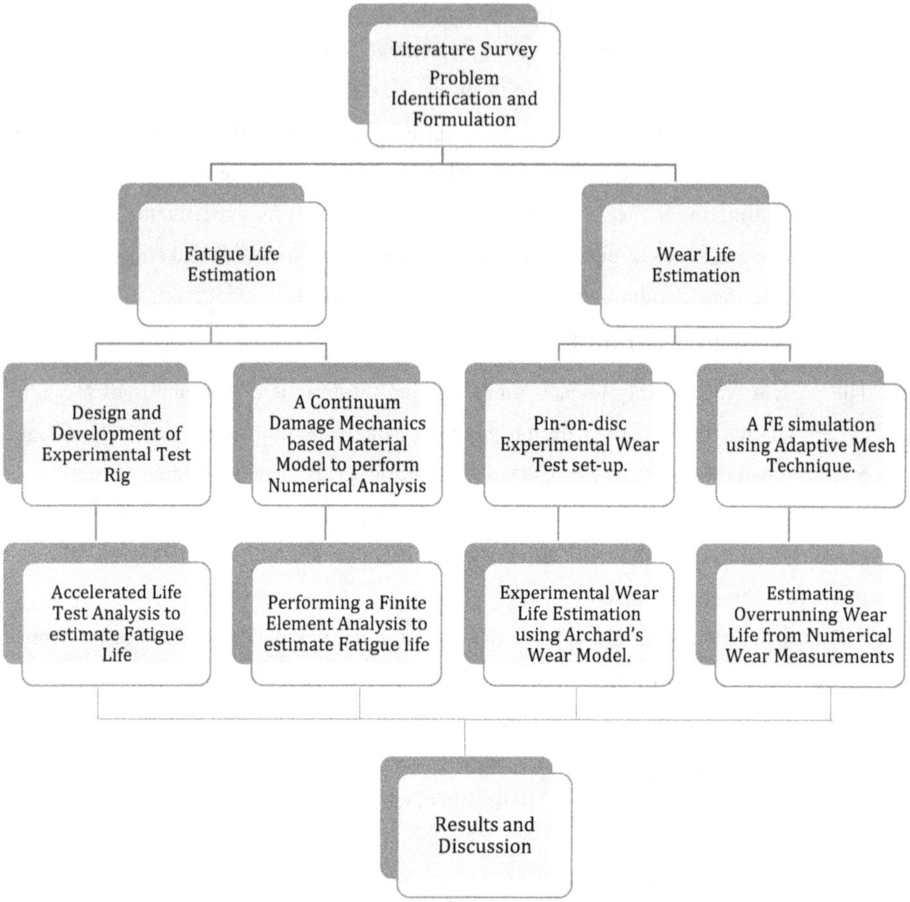

Figure 2.6 Methodology for the present work.

Furthermore, the fatigue life estimated using the experimental data is validated with the continuum damage mechanics-based numerical model, which better characterizes the observed fatigue damage under similar loading conditions. Similarly, the wear life of OWC is estimated for overrunning mode through pin-on-disc experiments using Archard's wear model and validated using an FE-based numerical model.

Chapter 3#

Design and Development of Experimental Test Rig for Fatigue Life Estimation of OWC

This chapter describes the selection, design, and assembly of the test rig components utilized to develop the experimental test rig for fatigue life estimation of OWC.

GRAPHICAL ABSTRACT

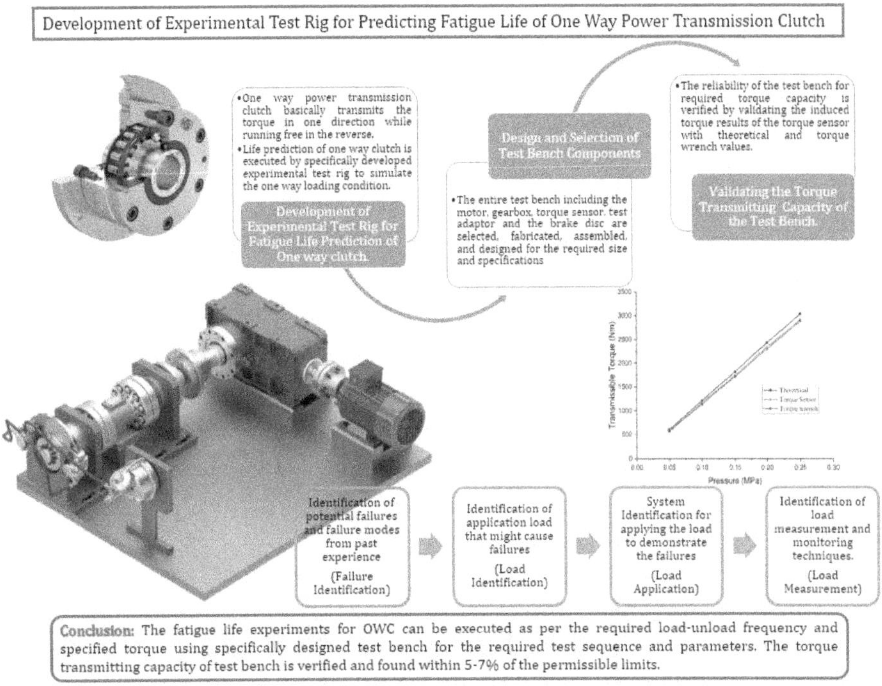

OWC exhibits high cycle fatigue during its lifespan under repeated loads. The conventional fatigue experiments for rolling element bearings have been carried out on a three-point bending experimental test rig, which comprises a rotating shaft with a pedestal bearing support on either side and a central load for the load reversals.

*#**Karan A. Dutt**, Shashikant J. Joshi, Dhaval B. Shah, Shashikant B. Soni, Deepak Prajapati. "Design and Development of Experimental Test Rig for Fatigue Life Estimation of One-way Power Transmission Clutch." International Conference. AIAMMS 2023 organized by JECRC University-Jaipur.*

In the case of OWC, the same test rig cannot perform the fatigue experiments since OWC is functionally different from bearing in terms of fatigue operations. It is, therefore, necessary to establish a specific test rig on which OWC can be easily operated as per their defined operating parameters, and fatigue tests can be conducted as per the required test sequences.

3.1 Conceptual Design of Fatigue Test Rig for OWC

As indicated in **Figure 3.1**, a flow diagram has been followed to develop the test rig for applying the required loading and operating conditions. The first step towards developing the fatigue test rig for OWC includes the identification of potential failures from field experience or through application feedback. Before evaluating the fatigue life of any component, it is required to identify the modes of failure endured by specimens. In the present case, the surface damage of the inner contact surface due to repeated loading-unloading is considered as critical damage state. Further, the corresponding load values to develop such a failure mode must be determined by qualitative tests. The theoretical design also provides valuable insights for evaluating the critical load for such material failures.

Later, the load application system is conceptualized and designed to obtain the defined failure mode using available electro-mechanical technologies. Lastly, the load measurement and monitoring system is established to measure and monitor the real-time loading patterns for performing the fatigue experiments.

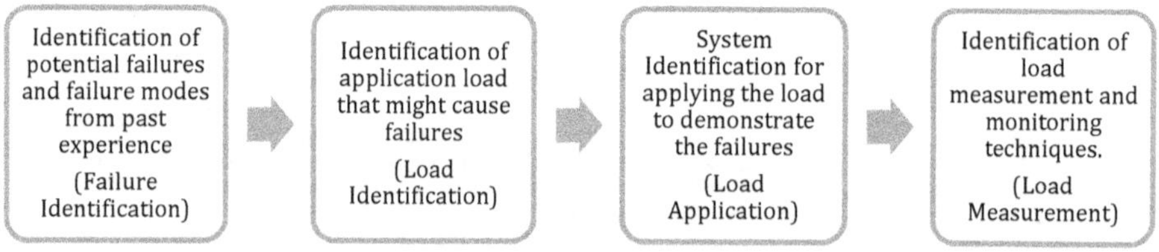

Figure 3.1. A flow diagram for developing a fatigue test rig.

As shown in **Figure 3.2**, a schematic line diagram presents the experimental setup for simulating the one-way power transmission function for performing fatigue experiments. The conceptual design of the test rig consists of three essential systems: a driving member, which acts as a load application system within the required range; a driven member, which acts as a load absorption system; and a mechanical torque sensor, which acts as a load measurement system. The driving member transmits the power from the inner race of the specimen to the outer race, which is coupled with the driven member. The driven member is restricted through a hydro-pneumatic brake system with the help of brake disc and hydraulic callipers. A rotary torque sensor measures the induced load on the specimen.

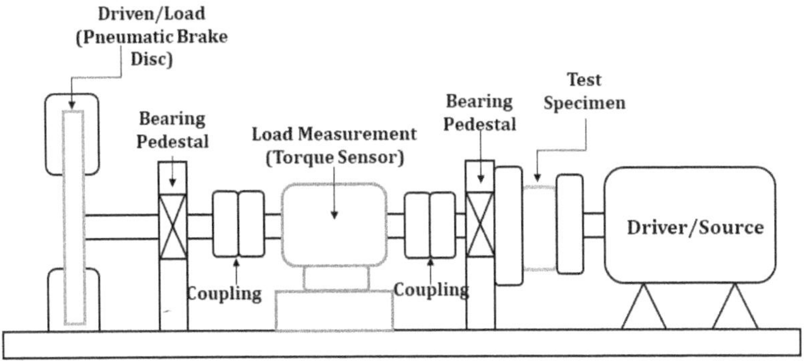

Figure 3.2. A line diagram of the proposed experimental test rig

3.2 Development of Experimental Test rig for Fatigue Life Estimation

An experimental test rig is designed, fabricated, and assembled as per the discussion in section 3.1 to simulate the required function of one-way torque transmission to estimate the engagement fatigue lives of OWC under 5000 N m of torque transmission capacity. As per the required torque application, the selection of the driving system is carried out, which comprises a 15 kW, 4-pole, 3-phase induction motor of 1500 rpm with a 3-stage helical type speed reducer of 54:1 reduction ratio, capable of delivering the maximum torque of 5000 N m at 28 rpm. The power is transmitted from the gearbox output shaft to the inner race of the OWC specimen. The clutch's outer race is coupled with the driven member through a torque-measuring device. The driven member is provided with the hydro-pneumatic brake system, which acts as a dynamometer to induce the required intensity of torque at the specimen. The test specimen is installed between driving and driven members with the help of an adjustable adaptor to accommodate the required size of the clutch. The hydro-pneumatic system consists of pressure measuring and pressuring regulating devices. The braking load is applied to the brake disc and measured using a rotary-type torque sensor of 5000 N m capacity equipped in line with the drive shaft and brake disc shaft through couplings with pedestal bearing supports. An electronic digitizer records the torque measurement values from the torque sensor. A remote input/output Variable Frequency Drive (VFD) controls the whole mechanical system to set the required loading-unloading frequency and apply speed variations. The pressure measuring gauges and regulatory devices like electronic pressure gauges, i.e. solenoid switches, are controlled by a Programmable Logic Control (PLC) module to set the required effective pressure.

3.3 Design and Selection of Test Rig Components

The bearings and couplings are selected as per the required geometrical specifications for developing the test rig. Four quantities of single-row deep groove ball bearings specified as SKF 6018 of size 90 x 140 x 24 **[141]** are utilized in pedestal supports and selected for the required test rig configuration. Similarly, two quantities of rubber bush couplings specified as RB-285-11 of size 48 x 285 x 227 **[142]** are selected as a coupling unit of the test rig. A torque measuring device specified as an MRT rotary torque sensor **[143]** of 5000 N m capacity is selected and utilized in the test rig.

3.3.1 Design of adaptor for OWC test specimen

The torque supplied from the gearbox output is transmitted to the OWC specimen through an adaptor. The adaptor assembly consists of several components fabricated using carbon steel EN9 material, which consists of an integrated flange shaft, spigot cover, socket plate, etc. The exploded view of the adaptor is shown in **Figure 3.3**.

The torque from the gearbox output flange is transmitted to the integrated flange shaft of the adaptor to the OWC inner race through the key. The outer race of the clutch is provided with a spigot cover to transfer the torque from the inner race to the socket plate, which is connected to the output flange shaft, where the torque is transmitted to the driven end through flexible couplings. A specimen adaptor for the available OWC sample is designed based on the maximum torque transmitting capacity. The integrated flange shaft diameter (d) is calculated based on strength using **Eq. (3.1) [144]** for a maximum torque capacity (T) of 5000 N m. The flange thickness is considered 0.5 times the shaft diameter [144],

$$T = \frac{\pi}{16} \times \tau \times d^3 \quad \textbf{(3.1)}$$

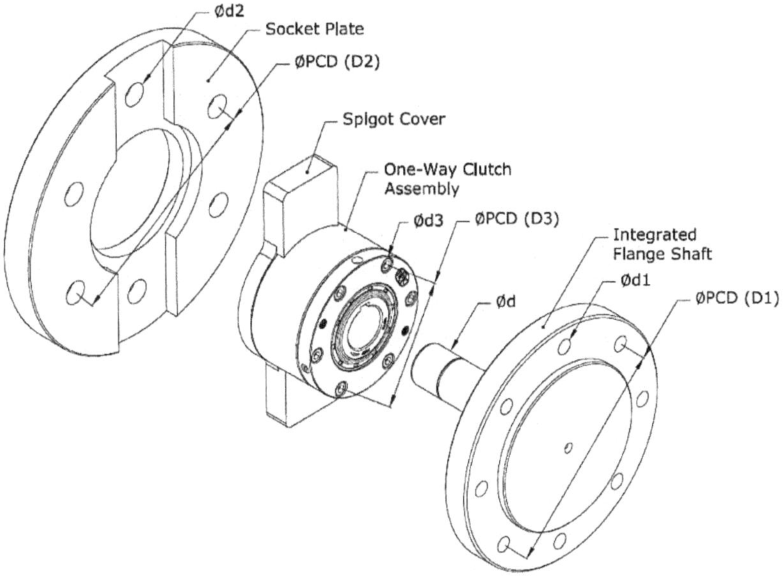

Figure 3.3. Exploded view of adaptor for OWC test specimen.

The clamping screws of the spigot cover, socket plate, and integrated flange shaft are manufactured from 5.8 grade BS 6104 alloy steel and selected using **Eq. (3.2) [144]** for maximum transmissible torque. The material properties for respective components are listed in **Table 3.1.**

$$T = n_s \times \frac{\pi}{4} \times Screw\ size^2 \times \tau_{screw} \times \frac{Pitch\ Circle\ Diameter}{2} \quad \textbf{(3.2)}$$

Table 3.1. Material properties of specimen adaptor.

Material Properties	EN9 Carbon Steel	BS 6104 Alloy Steel Clamp Screw
Ultimate tensile strength	600 MPa	520 MPa
Yield strength	310 MPa	380 MPa
Shear strength	178 MPa	210 MPa
Modulus of Elasticity	210 GPa	-

The specimen adaptor components designed using **Eq. (3.1-3.2)** for required geometrical dimensions against the required transmitting torque of 5000 N m are listed in **Table 3.2**.

Table 3.2. OWC adaptor design parameters.

Parameter	Value
Integrated flange shaft diameter, d	56 mm
Integrated flange pitch circle diameter, D_1	150 mm
Integrated flange clamp screw size, d_1	10 mm
Socket plate pitch circle diameter, D_2	165 mm
Socket plate clamp screw size, d_2	12 mm
Spigot cover pitch circle diameter, D_3	112.5 mm
Spigot cover clamp screw size, d_3	10 mm
Integrated flange thickness, t_1	28 mm
Socket plate thickness, t_2	55 mm
Spigot cover thickness, t_3	24 mm

3.3.2 Design for brake disc

Two quantities of calliper brakes constructed from FG 260 material are utilized for the test rig for the required torque capacity, and the technical characteristics of Brembo callipers **[145]** are described in **Table 3.3**. The hydro-pneumatic brake system consists of a hydraulic booster with a piston and braking pad arrangement that transfers the pneumatic pressure into hydraulic pressure. The moving piston transfers the actuating pressure to the braking pad pressed against the revolving brake disc to generate the required amount of torque. The actuating pressure is ten times the pneumatic pressure induced by the compressed air through a solenoid control valve to control the air pressure supplied to the booster. The typical brake disc arrangement utilized for required torque generation in the test rig is shown in **Figure 3.4.** The available theoretical torque capacity is calculated using **Eq. (3.3) [144],**

T = μ x normal force x Effective Radius = μ x (hydraulic pressure x surface area) x r_{ef} = μ x (boosting factor x pneumatic pressure) x (no. of brakes x no. of pads per brake x pad area) x r_{ef}

$$T = \mu \times 10 \times p \times 2 \times z_p \times a_p \times r_{ef} \quad \textbf{(3.3)}$$

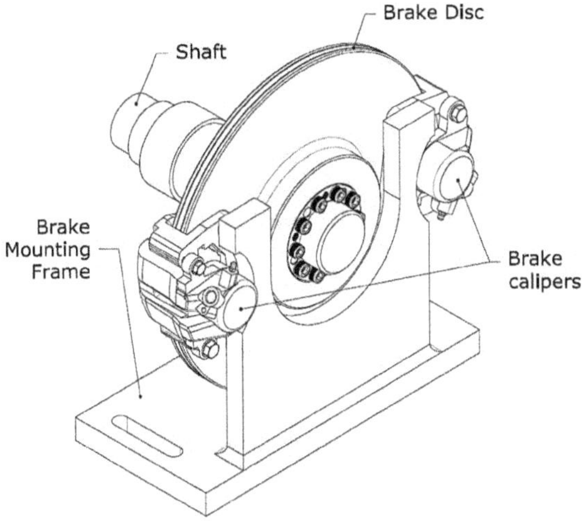

Figure 3.4. Hydro-pneumatically operated disc brake assembly.

Table 3.3. Technical characteristics: Brembo caliper brake **[145]**.

Parameter	**Value**
No. of piston, n_p	2
Piston diameter, d_p	60 mm
Maximum operating pressure, $\boldsymbol{p_{max}}$	15 MPa
No. of pads, z_p	2
Disc diameter D, Minimum d	284 mm
Maximum D	410 mm
Disc thickness at Min. and Max. Worn	16-14.5 mm
Weight with pads	9.35 kg
Effective rolling radius, r_{eff}	(D/2)-31 = 173.5 mm
Co-efficient of friction	0.35
Brake fluid	dis.20.2846.11/51(A)
Hydraulic oil	dis.20.2846.31/71(A)
Pad area, a_p	5280 mm^2

3.4 Experimental Test Rig for Fatigue Analysis of OWC

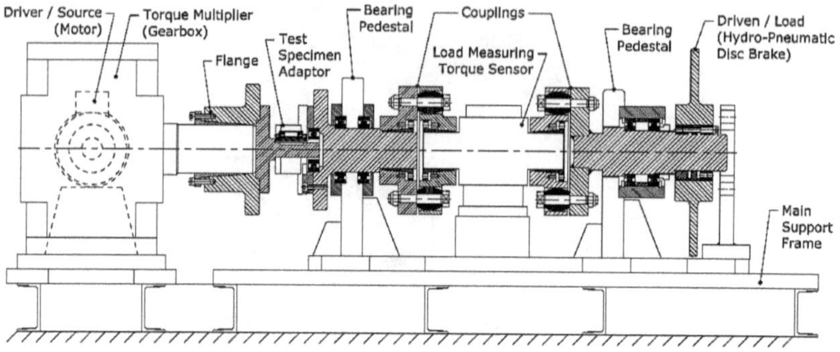

(a)

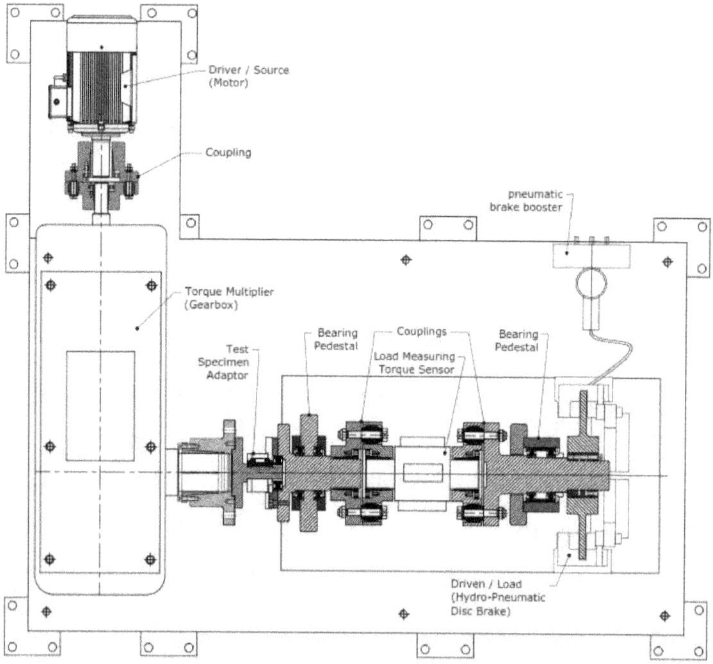

(b)

Figure 3.5. A general arrangement of the experimental test rig: **(a)** Front view **(b)** Top view

The general arrangement of the entire test rig in terms of 2-dimensional front view and top views is displayed in **Figure 3.5**. The parametric design of the assembly components, as described in section 3.2, is also illustrated in terms of 3-dimensional solid CAD assembly, as shown in **Figure 3.6.**

Figure 3.6. A 3-dimensional solid model of the experimental test rig.

Later, the experimental test rig for fatigue life estimation of OWC is fabricated with the specified components and assembled, as displayed in **Figure 3.7.**

Figure 3.7. An experimental test rig developed to simulate the one-way power transmission function to estimate the fatigue life of OWC.

A test program, as displayed in **Figure 3.8,** is prepared to run the test rig with the required test inputs and implement the necessary torque adjustments by controlling the air pressure using a pressure regulator. As per the assigned value of air pressure, the pressure regulator, i.e., the solenoid control valve, accurately controls the airflow to operate the hydraulic pads of the calliper through a hydraulic-pneumatic booster, thereby inducing the required torque at the brake disc. The program executes the specific instructions in terms of parameters such as the direction of loading, motor speed and loading-unloading frequency assigned through remote input/output variable frequency drive for the specified time intervals with the help of a programmable logic controller module. This combined control of parameters exhibits the functional requirement for OWC application. The program continues to record the number of cycles for a predefined loading-unloading frequency until torque loss at the specimen due to the failure, which is measured by the torque sensor.

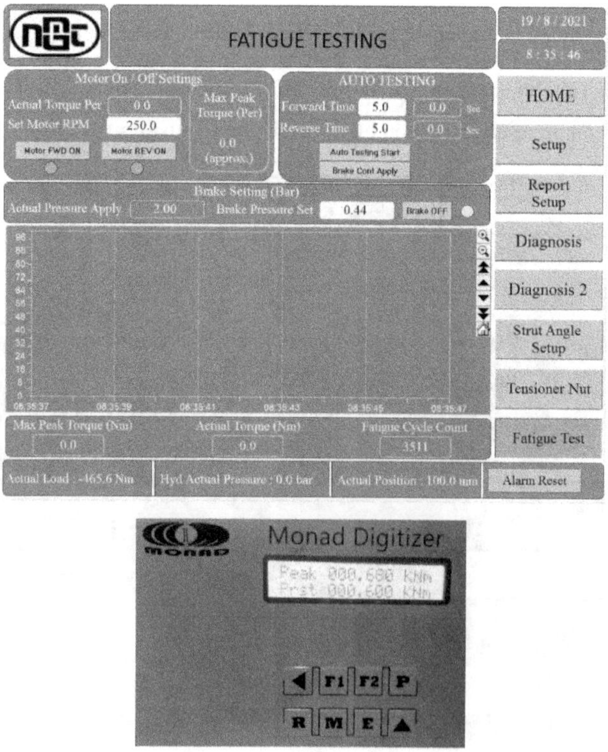

Figure 3.8. A PLC-based GUI dashboard for monitoring and controlling test parameters with the digitizer.

3.5 Test Rig Operational Validation for Torque Transmitting Capacity

In order to produce effective endurance tests of one-way clutch, it is required to know whether the developed test rig can deliver its designed torque transmitting capacity. For this purpose, the torque transmitting capacity of the test rig is validated by measuring the actual torque induced at the specimen.

The maximum transmissible torque induced at the test sample is measured by a rotary torque sensor installed between the driving and the driven member. The torque transmitting capacity of the test rig is initially verified by performing a validation test in load condition, which is executed by rotating the brake disc using a hand-operated torque wrench at the driven end. The same has been recorded by the torque sensor and verified with the theoretical value calculated using **Eq. (3.3).** The results are listed in **Table 3.4.** The graphical representation of transmissible torque with respect to pneumatic pressure is depicted in **Figure 3.9.**

Table 3.4. Torque measurement and validation data of test rig.

Pressure, MPa	Torque measured by torque sensor (As shown in digitizer), N m	Torque measured by torque wrench at brake disc, N m	Theoretical torque at brake disc, N m	% Difference from theoretical	
				Torque sensor	Torque wrench
0.05	588	569	607.3	3.2	6.3
0.1	1167	1140	1214.5	3.9	6.1
0.15	1741	1715	1821.8	4.4	5.9
0.2	2342	2298	2429	3.6	5.4
0.25	2917	2888	3036.3	3.9	4.9

The validation of transmissible torque for the test rig is carried out using a torque wrench and torque measuring device, i.e., a torque sensor. The reported torque wrench values have an approximate difference of 5 to 6% from the theoretical ones. On the other hand, the torque sensor follows quite closer results to the theoretical values, with around 3 to 4 % of the difference. The deviation in the torque wrench values from theoretical values is more than the respective torque sensor values. It is because the torque sensor measures the torque values nearer to the unit under test than the torque wrench measurements, which have been carried out at the end of the brake disc. Thus, the torque sensor provides accurate results, which are also within the standard error criteria of the instrument, i.e. ±5% of theoretical values. Such analysis conforms to the compatibility of the test adaptor, which is designed to comprehend various-sized clutches on the same test rig for fatigue life estimation. It also ensures the accuracy of the load application and the accuracy of conducting the fatigue test with specified torque capacities on the test rig. The fatigue data obtained from these experiments can be further processed to estimate the life of OWC at operating levels.

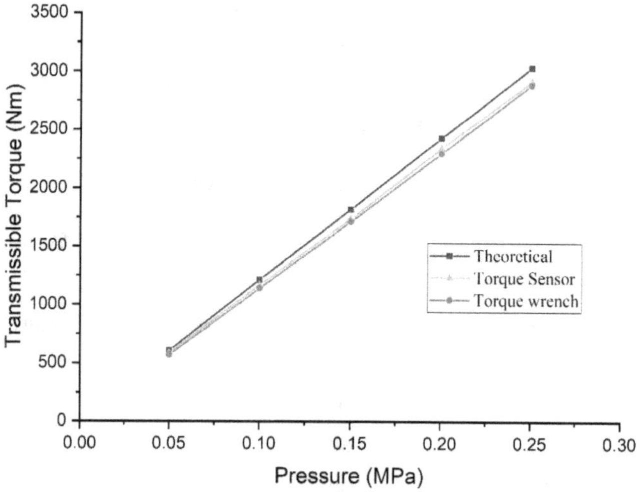

Figure 3.9. Validation of torque capacity of the test rig.

The high cycle fatigue of OWC is thus examined using this test rig specifically developed to accurately capture the required one-way function, which was not carried out earlier in the available literature. The fatigue life comparison for different-sized clutches on the same test rig is executed and analysed to understand the effects of clamping zone size on fatigue lives. The experimental fatigue life results are analysed, and the load-life relationship for OWCs is developed using these data. Numerical analysis can also be carried out to estimate the load-life relationship of the clutch, which can be validated with the experimental results.

3.6 Summary

This chapter presented the design concept and the development of a test rig to perform fatigue experiments on OWC. The test rig is designed and developed using a drive member, a load member, a torque measuring device, and a specific adaptor to accommodate the required size of the test sample. The selection and design of the test rig components for the required torque capacity have been carried out using applicable machine design criteria. The torque at which the experiments are required to be performed is exhibited using a hydro-pneumatic disc brake system. The test rig, including the test adaptor and the brake disc, is selected, fabricated, assembled, and designed for the required size and specifications. The assembly parts are designed to meet the required torque transmitting capacity. The reliability of the test rig for the required torque capacity is verified by validating the induced torque results of the torque sensor with theoretical and torque wrench values, which are within the acceptable limits. The fatigue life experiments for OWC can be executed as per the required load-unload frequency and specified torque using this test rig for the required test sequences and test parameters.

Chapter 4#

Experimental Test Plan and Test Methodology for Fatigue Life Estimation of OWC

This chapter deals with the experimental test plan and test methodology utilized for performing the fatigue experiments on test samples. The effect of various parameters on fatigue life is also discussed. The load-life relationship is developed using accelerated tests and compared with the conventional fatigue model. The failure analysis of fractured specimens is also discussed.

GRAPHICAL ABSTRACT

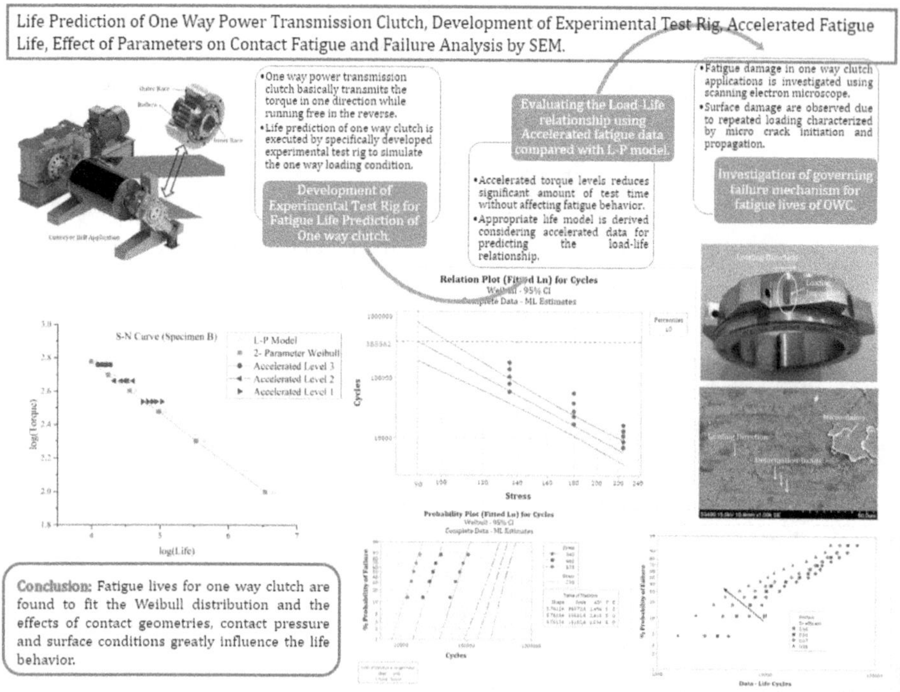

*#**Karan A. Dutt,** Shashikant J. Joshi, Dhaval B. Shah, Shashikant B. Soni, Deepak Prajapati (2024). "Experimental investigation of fatigue for one-way power transmission clutches using accelerated life tests." Results in Engineering. Volume 22, 2024, 102136, ISSN 2590-1230.*

4.1 Fatigue Test Methodology

The fatigue life of OWC generally falls into the category of HCF. Since the objective of the present work is to predict fatigue life at use conditions, the test specimens are required to be tested at values higher than the specified values to accelerate the failure mechanism and save a considerable amount of testing time. In order to track the failure of the components, a permissible torque limit of 5% reduction has been decided. That means when the torque sensor measures the induced torque below 5% of the specified torque, a program directs the whole system to terminate the test cycle, and the clutch is considered failed.

4.1.1 Accelerated life tests analysis (ALTA)

Conventional life-data analysis utilizes times-to-failure data collected under usual operating conditions to quantify the life characteristics of the components. For many reasons, capturing such life data may be difficult or near impossible due to their long life, as in the present case of OWCs. It is vital for any structural element undergoing HCF to accelerate the test in laboratory conditions to acquire the results in a reasonable time. Accelerated life tests are performed to get information about the failure time distribution of products for such HCF. The tests are conducted in more severe conditions than the normal operating ones at which the components are supposed to operate to accelerate the failure, thereby shortening the life of the components. The most common types of accelerated tests include high usage rates and overstress loadings **[84]**. The high usage rate accelerates the usage life in the sense of running the product more than its actual usage by compressing the usage time.

On the other hand, overstress testing consists of running the product at a level higher than the normal stress level to shorten its life span by degrading its performance faster. Overstress testing is the most commonly used accelerating test method, and the same concept is being carried out in the current analysis. Accelerated life testing is further classified as qualitative test and quantitative test.

In qualitative tests, the main focus will be identifying the modes of failure without making any prediction of life. In quantitative tests, the life prediction at use condition is the prime concern from the data obtained in accelerating tests. The block diagram, as shown in **Figure 4.1,** depicts the test plan carried out in the present work in accelerating life testing for one-way freewheels.

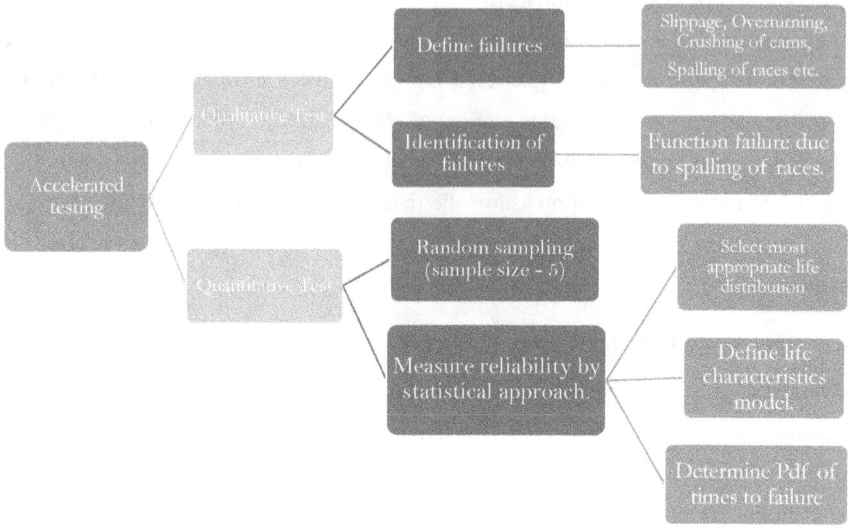

Figure 4.1. Accelerating test methodology utilized in the current analysis **[84]**

The potential failure modes are identified first without making any life prediction, and the life prediction is carried out at normal operating conditions based on the data obtained in accelerating tests by applying appropriate underlying stress life models **[84, 146].** These models incorporate the relationships accounting for the effect of stress (load) values on the life of the components under study. The stress loading in the accelerated test can be constant, cyclic, step or progressive levels. The present discussion is limited to a single variable cyclic load where one-way freewheel specimens undergo repeated loads of similar stress patterns. **Figure 4.2** depicts a sinusoidal pattern of cyclic loading for the accelerated test of a specimen.

The life data are recorded for individual load values on the test rig for the specified set of parameters. One torque value corresponds to one specimen to be tested; thus, for each elevated torque value, new specimens are required to be tested, and respective fatigue data are recorded. Also, the experiments are performed on five samples for each accelerating torque value to ascertain the repeatability of the results. A test matrix for conducting experiments has been prepared. After undergoing the failure for respective elevated torque values, the new specimen replaces the previous one through an adjustable adaptor for continuing further experiments. The accelerating load values beyond the use value are identified and tested for the number of cycles to failure. The potential failures are identified, and the life cycle values are recorded for individual load values.

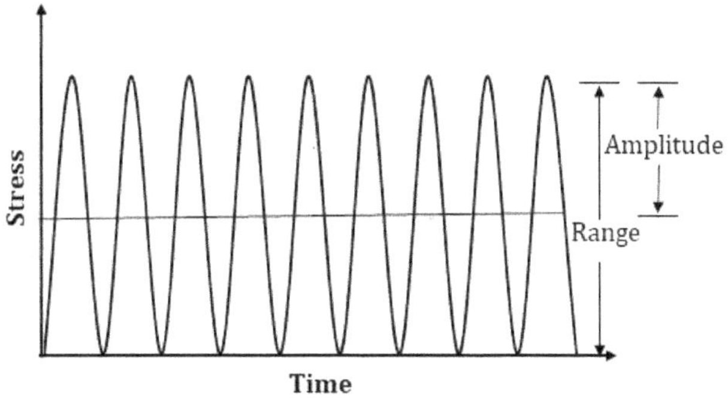

Figure 4.2. Cyclic loading patterns for fatigue testing of OWC.

4.1.2 Selection of elevated load levels and load-life relationship

In order to estimate the fatigue life of OWC, one load cycle must be considered as one engaged rotation in the load direction along with reverse rotation, which is just sufficient enough to disengage in the opposite direction. The stress loading relevant to one-way clutches is a constant load in one direction and no load in reverse. Elevated torque levels are selected in such a way as to accelerate only the modes of failure without introducing any other effect which would never occur during normal operating conditions **[147-148]**. Apparently, increasing stress levels would decrease test time. However, as the stress levels move farther away from the operating condition, the failure mechanism may change, and the uncertainty in extrapolation may also increase. Therefore, it is necessary to be careful while selecting the elevated load levels. Generally, these torque levels should fall outside the specification limits of the specimen but inside the design limits, as depicted in **Figure 4.3 [84]**. If the values are unknown, qualitative tests should be performed to confirm these torque levels. In the current study, the specimens are designed for 2.5 times the nominal specified torque value. Qualitative tests are performed to confirm these torque levels. In addition to the selection of torque levels, the torque application must also be accomplished in quite a logical, controllable and quantifiable fashion. The physical understanding of the failure mechanism can be justified by an empirical model extrapolated based on the extensive experience of how these failure mechanisms interact with accelerating variables. The fundamental relationship of fatigue life as a function of torque is modelled using the appropriate law of physics depending on the failure mechanism.

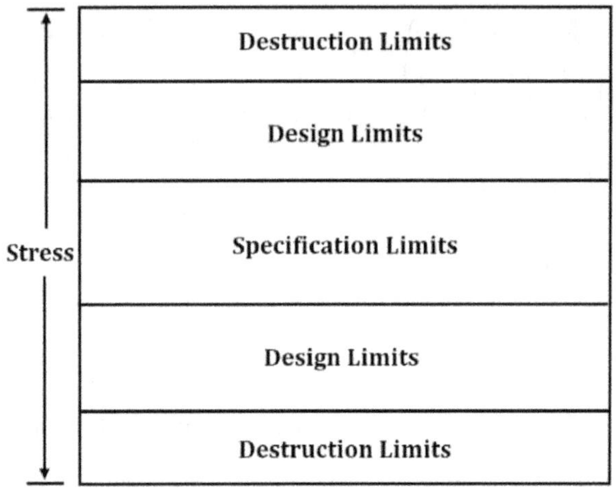

Figure 4.3. Selection of elevated load levels **[84]**

The first step in executing ALTA is to select an appropriate life distribution that can exhibit specific trends in failure rates. It could be any life measure in the form of mean, median, reliability function or failure rate. The most common life distributions include exponential, Weibull and Log-normal life distributions, whereas exponential life distribution is widely used for its simplicity. Conversely, Weibull and log-normal life distributions involve more complex calculations to define failure rates and are most appropriate for long-life product applications **[84]**.

The next step is to select a load-life relationship or to define a model that can describe the life characteristics of the data from one load level to another. There are infinite ways to project a specific point from elevated stress level to operating stress level to define a relationship, as depicted in **Figure 4.4 (a).** It is assumed that a function or a model can be mapped from the point of elevated stress level to operating stress level and can be described by a mathematical equation. Generally, two higher stress levels are required to properly project the function to an operating stress level **[84]**, as depicted in **Figure 4.4 (b)**. The more accelerating stress levels, the more accurately the life at operating stress levels can be predicted. Many stress-life models are available and utilized based on their applicability to specific applications, such as the Arrhenius model, which is well-known for its suitability to temperature-dependent ALTA **[84]**. In the same way, the inverse power model is found to be an appropriate and frequently used load-life relationship for contact elements to describe the effect that the load has on life, derived by Meeker and Escobar **[86]**.

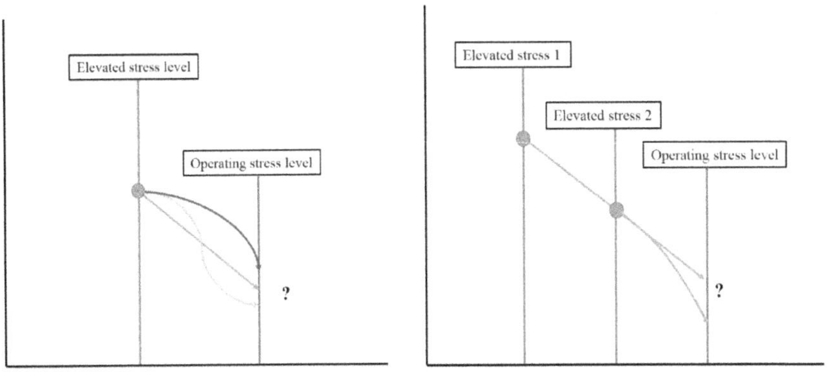

(a) **(b)**

Figure 4.4. Mapping from elevated stress level to operating stress level for **(a)** a single elevated stress level, **(b)** two elevated stress levels

It is noted that within the framework of accelerated analysis, certain assumptions are made: (i) the failure modes at higher torque levels remain unchanged, (ii) higher torques speed up the prevailing failure mode to normal stress, and (iii) the exact statistical characterization needs to be valid at the higher and operating torque levels to analyse the results.

4.2 Test Plan and Methodology

The selection of a good test plan is necessary to acquire the desired information without wasting time, money and effort on testing. Performing accelerated tests could sometimes prove to be a difficult task and an expensive endeavour. Preparing a test plan before conducting the accelerated tests is always advisable to minimize the costs and test duration. It also helps to estimate the reliability of the available operating conditions accurately. Such test plan must be used to decide the appropriate torque levels at which the ALTA should be carried out and the number of test specimens that must be allocated at these torque levels.

The highest torque levels can be determined based on engineering knowledge, qualitative tests, and experience with similar products. The quality of an effective test plan depends on the magnitude of stress levels and their extrapolation to the operating level. It is evident that higher stress levels will help to end testing faster than lower stress levels, but it may violate the assumed life distribution and load-life relationship. Higher magnitude of accelerated stress levels correspond to quicker and more failures, further leading to better estimation efficiency but at the same time requiring more extrapolation to operating levels. Therefore, the trade-off between extrapolation and efficiency must be considered. Test plans with more stress levels are more sophisticated than those with fewer levels since they depend less on the load-life relationship validity assumptions. However, test plans with fewer levels are more convenient **[84].**

In the current analysis, three torque levels in equal steps are considered for collecting the life data at higher torque levels and extrapolating these data at the operating torque level to estimate the life. The number of specimens at each torque level is taken as five and run at the constant stress (torque) level. It means a total 15 number of fatigue tests are required to be conducted to map the accelerated model prominently. A roller OWC under investigation is categorized based on its torque transmitting capacities as light, medium, and high levels, identified throughout this study as specimens A, B, and C, respectively. The ALTA is carried out for all three specimens under their elevated load levels. Ultimately, 45 fatigue experiments are executed under various loading conditions for different sizes of OWC.

An experimental test matrix is prepared as displayed in **Table 4.1** to perform the experiments effectively and record the life cycle data of three OWC specimens categorized in light, medium and heavy-duty capacities, i.e. specimen A, specimen B and specimen C, respectively, to further estimate the life of the clutch. Five specimens from each category are randomly selected for the accelerated test under each elevated torque level.

Table 4.1 An experimental test matrix for conducting fatigue experiments.

NC 30 (Low torque capacity)				
	Use Condition	Higher stress conditions		
Torque	**90 N m**	135 (1.5 x UC)	180 (2 x UC)	225(2.5 x UC)
Sample unit	Life in cycles (Load counts)			
1				
2				
3				
4				
5				
NC 50 (Medium torque capacity)				
	Use Condition	Higher stress conditions		
Torque	**230 N m**	345 (1.5 x UC)	460 (2 x UC)	575(2.5 x UC)
Sample unit	Life in cycles (Load counts)			
1				
2				
3				
4				
5				
NC 80 (High torque capacity)				
	Use Condition	Higher stress conditions		
Torque	**1063 N m**	1595 (1.5 x UC)	2126 (2 x UC)	2658(2.5 x UC)
Sample unit	Life in cycles (Load counts)			
1				
2				
3				
4				
5				

4.3 Results and Discussion

All three specimens are investigated under ALTA as per the test matrix displayed in **Table 4.1** on a fatigue test rig. The test parameters are specified for each torque value and are controlled analogously to the required output. The loading frequency within the operating range of the control system is set as 0.2 Hz to complete the experiment in the stipulated time frame. Similarly, the practical engagement speed is defined as 30 rpm. The disengagement speed is taken as 120 rpm to exhibit proper disengagement. **Table 4.2** indicates the values of input parameters required to achieve corresponding elevated torque values for experiments.

Table 4.2 Accelerating test parameters.

Test specimen	**Operating torque, N m (Use condition)**	**Accelerated torque, N m (Pneumatic pressure, N/mm^2)**		
Specimen A (Small)	90	135	180	225
		(0.0105)	(0.0140)	(0.0176)
Specimen B (Medium)	230	345	460	575
		(0.027)	(0.036)	(0.045)
Specimen C (Large)	1063	1595	2126	2658
		(0.1249)	(0.1664)	(0.208)

The experimental fatigue life data obtained for all three specimens under each elevated torque value for the defied test parameters are listed in **Table (4.3-4.5)**. These data are further required to be investigated to predict fatigue life at operating conditions using the test methodology discussed in section 4.2.

Table 4.3 Experimental results for life cycle analysis (Specimen A)

Use (Operating) torque level	**Accelerating torque levels, (Torque) N m**		
90	**135**	**180**	**225**
Sample unit	**Life cycle counts, N**		
1	58584	17145	7154
2	79858	22987	8789
3	104582	27845	10874
4	138958	38265	13215
5	177854	55874	16325

Table 4.4. Experimental results for life cycle analysis (Specimen B)

Use (Operating) torque level	Accelerating torque levels, (Torque) N m		
230	**345**	**460**	**575**
Sample unit	**Life cycle counts, N**		
1	105632	33568	18965
2	87548	40536	16584
3	55965	21420	12365
4	77854	26875	15254
5	66852	30658	14125

Table 4.5. Experimental results for life cycle analysis (Specimen C)

Use (Operating) torque level	Accelerating torque levels, (Torque) N m		
1063	**1595**	**2126**	**2658**
Sample unit	**Life cycle counts, N**		
1	38745	8965	5896
2	44698	11568	4258
3	28547	7865	3265
4	23658	12568	6258
5	66852	30658	14125

4.3.1 Life prediction using an accelerated model

The statistical analysis of accelerated life data includes the selection of the appropriate model that best fits the collected data and assessing its adequacy of fit. The analysis technique to be followed when assessing such accelerated life data is discussed in the following paragraph.

A statistical model for an accelerating life test comprises two important characteristics which must be defined and analysed. The first is a life distribution that displays the scatter in product life, and the second is the relationship between stress and life. As discussed in section 4.1, the life data analysis collected from the accelerated tests begins with selecting an appropriate life distribution. The scatter in fatigue life data is incorporated into the stress life relationship by selecting an appropriate life distribution. Due to its simplicity, the exponential distribution has been widely used as an underlying life distribution. There is very little chance of it being appropriate, and therefore, for most of the applications, lognormal and Weibull distributions are considered more appropriate life distributions due to the more involved calculations **[84]**. In the Weibull distribution models, the failure rates either increase or decrease in a simpler and more specific way. It distributes various product properties such as strength, resistance or elongation in accelerating tests. Sadeghi et al. **[27]** have revealed the suitability of using Weibull life distribution for analysing the contact fatigue data for bearings and thereby explicitly considered for OWC analysis in the present work. A 2-parameter Weibull distribution is expressed by **Eq. (4.1) [149]**

$$F(N;\ \beta, e) = 1 - exp\left[-\left(\frac{N}{\beta}\right)^{e}\right], N > 0 \tag{4.1}$$

Where F is the probability that the life is less than N cycles, while $\beta > 0,\ e > 0$ are two parameters, i.e. scale parameter and shape parameter, respectively, for the distribution. The shape parameter 'e' determines the shape of the distribution and represents the slope of the Weibull plot. The scale parameter 'β' determines the spread of the distribution and represents characteristic life when 63.2 per cent of the product sample population fails to operate their function. The two parameters are obtained using a maximum likelihood estimate **[150-151]**. Nelson and Lawless **[84, 150]**, among others, introduce the three-parameter Weibull distribution with additional location or threshold parameter. The threshold parameter presents the failure-free life of the component up to which, in any condition, the product continues to perform its function without any failure. It can be defined for individual life analysis but is rarely used for accelerated testing.

The fatigue life variations between different material domains mainly occur due to topological microstructure disorder. The scatter in fatigue lives recorded experimentally in identical OWC specimens is investigated. The analysis is carried out for predefined elevated torque values using the Minitab statistical software package **[152]**. **Figure 4.5** shows two-parameter Weibull probability plots for each accelerating torque level obtained for all three specimens from the experiments. The prediction of fatigue life at operating level torque (230 N m) for all relevant specimens is estimated from these accelerating values by considering the inverse power law relationship, as shown in **Figure 4.6.** The statistics table includes the shape parameter, which defines the failure rate with respect to time, and the scale parameter, which states the mean time to failure. The value of the scale parameter indicates the number of fatigue cycles for having 63.2% of components failed. The small Anderson-Darling statistic value shows the goodness of fit. The corresponding Weibull statistics are listed in **Table (4.6-4.8)**. The Weibull relation plot describes the linear relationship estimated from accelerated life test analysis, as shown in **Figure 4.7** for all three specimens. It shows that the inverse power law life-stress characteristics fit pretty well with experimentally observed data from the Weibull analysis and the life of OWC under operating conditions is predicted using an accelerating model.

Table 4.6. Weibull statistics for accelerating life test analysis

(Specimen A, $\alpha = 3.2$, $\mu = 0.05$)

Two-parameter Weibull estimation				
Torque levels, N m		**Weibull slope, e**	**Characteristic life, β (cycles)**	**L_{10} life (cycles)**
(Accelerating)	225	3.83	12491	6941
	180	2.57	36680	15281
	135	2.90	126064	58021
(Operating)	90	2.92	831162	385562

Table 4.7. Weibull statistics for accelerating life test analysis

(Specimen B, $\alpha = 3.2$, $\mu = 0.05$)

Two-parameter Weibull estimation				
Torque levels, N m		**Weibull slope, e**	**Characteristic life, β (cycles)**	**L_{10} life (cycles)**
(Accelerating)	575	7.53	16442.8	12195
	460	5.31	33233.1	21753
	345	5.06	85743.8	39504
(Operating)	230	5.76	328192	222052

Table 4.8. Weibull statistics for accelerating life test analysis

(Specimen C, $\alpha = 3.2$, $\mu = 0.05$)

Two-parameter Weibull estimation				
Torque levels, N m		**Weibull slope, e**	**Characteristic life, β (cycles)**	**L_{10} life (cycles)**
	2658	5.66	5196.4	3658
(Accelerating)	2126	7.01	12155.1	7967
	1595	5.46	36354	24941
(Operating)	1063	5.39	170274	117074

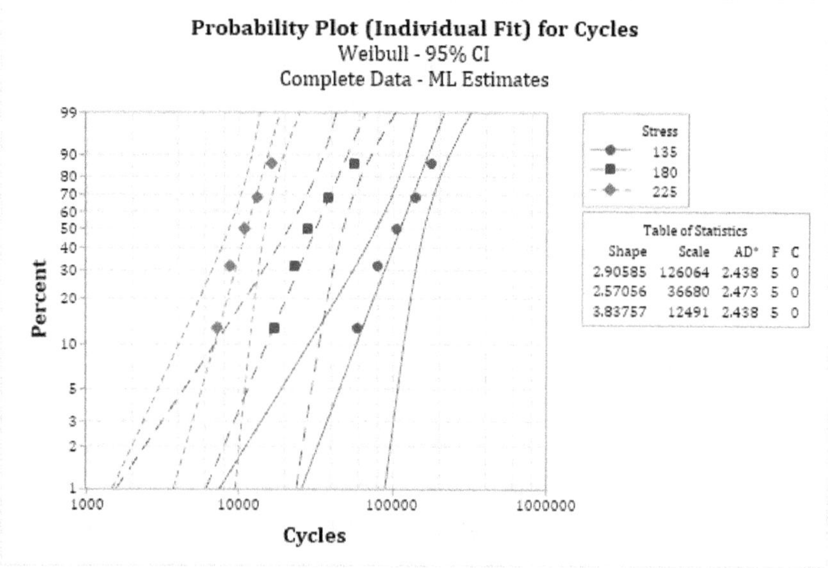

(a)

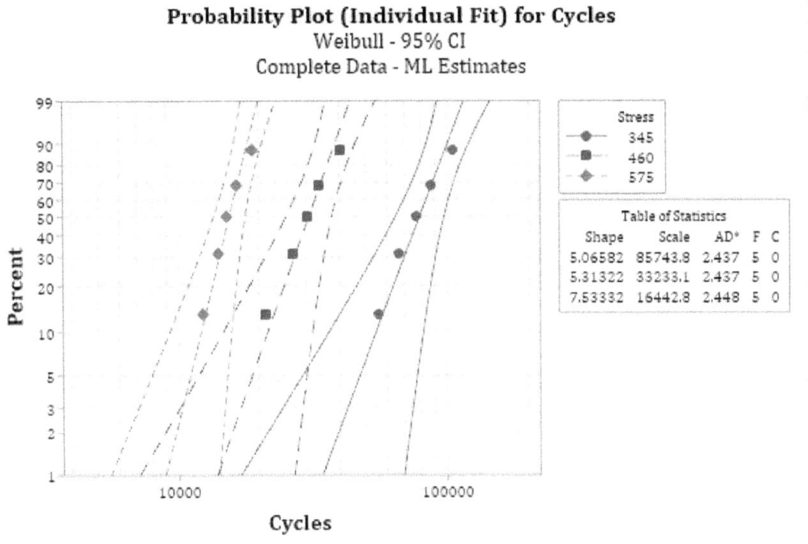

(b)

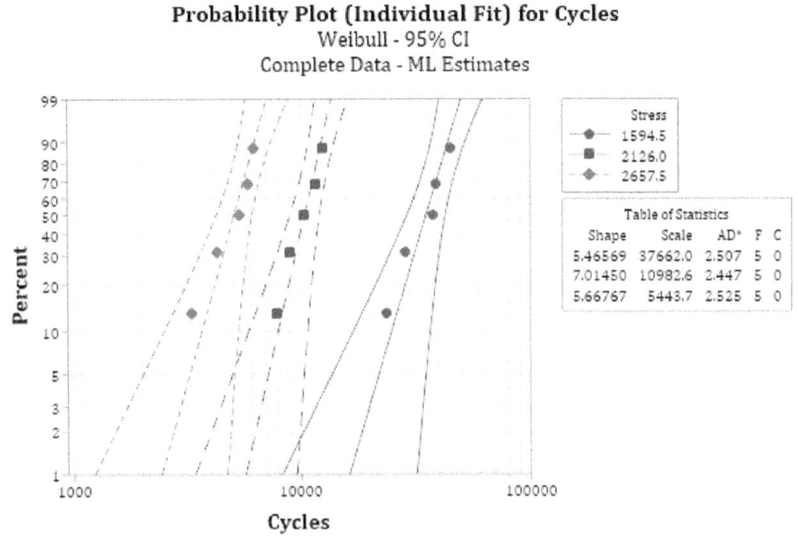

(c)

Figure 4.5. Weibull probability plot for accelerating torque values for **(a)** Specimen A, **(b)** Specimen B, **(c)** Specimen C

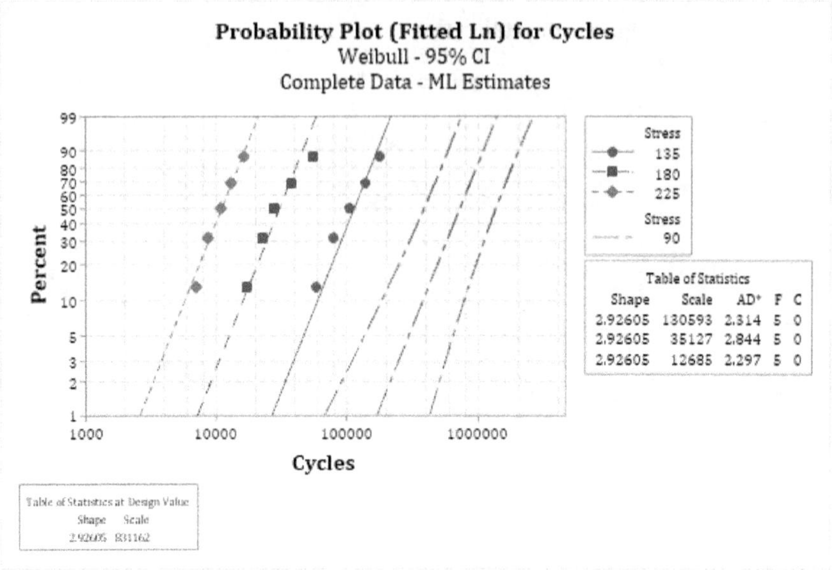

Probability Plot (Fitted Ln) for Cycles
Weibull - 95% CI
Complete Data - ML Estimates
Percent
Cycles
Stress
135
180
225
Stress
90
Table of Statistics
Shape Scale AD* F C
2.92605 130593 2.314 5 0
2.92605 35127 2.844 5 0
2.92605 12685 2.297 5 0
Table of Statistics at Design Value
Shape Scale
2.92605 831162

(a)

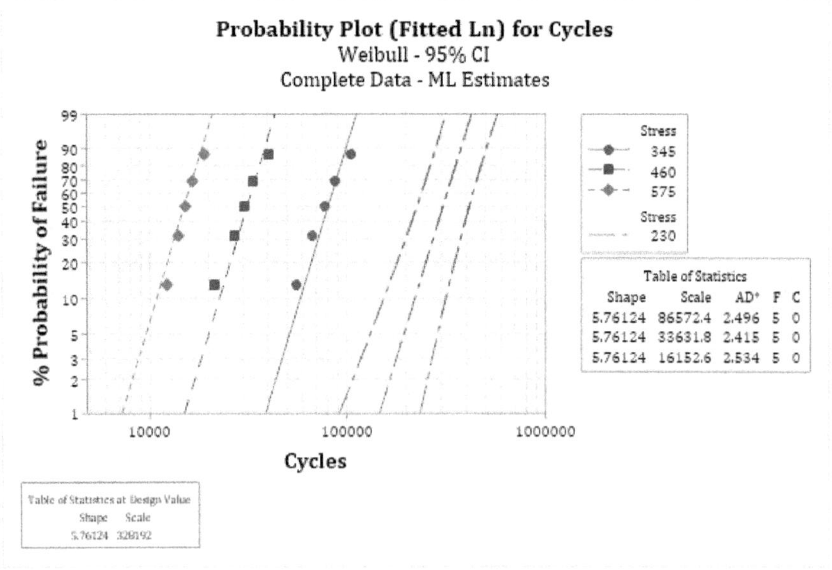

Probability Plot (Fitted Ln) for Cycles
Weibull - 95% CI
Complete Data - ML Estimates
% Probability of Failure
Cycles
Stress
345
460
575
Stress
230
Table of Statistics
Shape Scale AD* F C
5.76124 86572.4 2.496 5 0
5.76124 33631.8 2.415 5 0
5.76124 16152.6 2.534 5 0
Table of Statistics at Design Value
Shape Scale
5.76124 328192

(b)

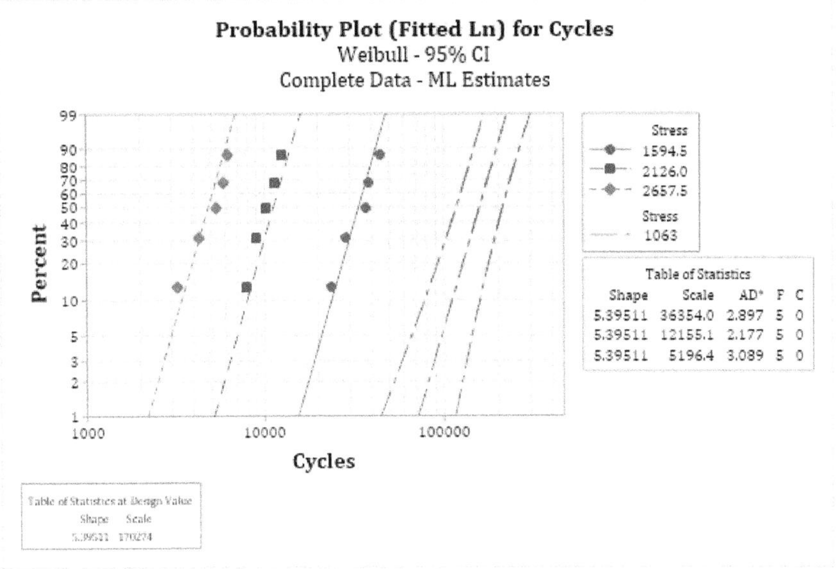

(c)

Figure 4.6 (a). Weibull probability plot fitted to operating torque value for **(a)** Specimen A, **(b)** Specimen B, and **(c)** Specimen C using inverse power law relationship.

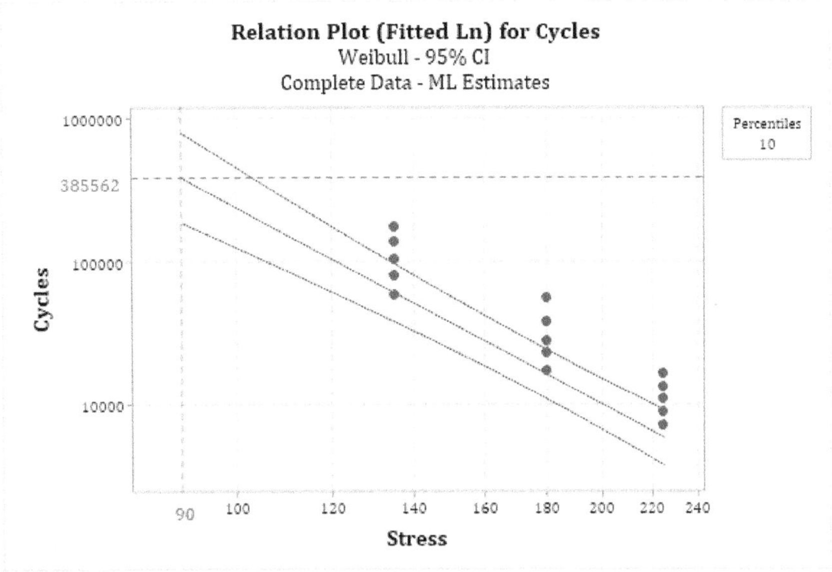

(a)

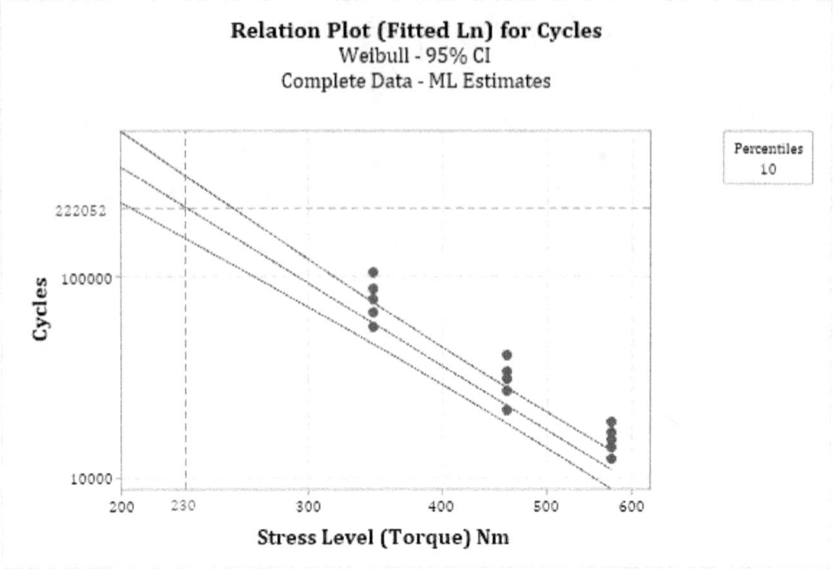

(b)

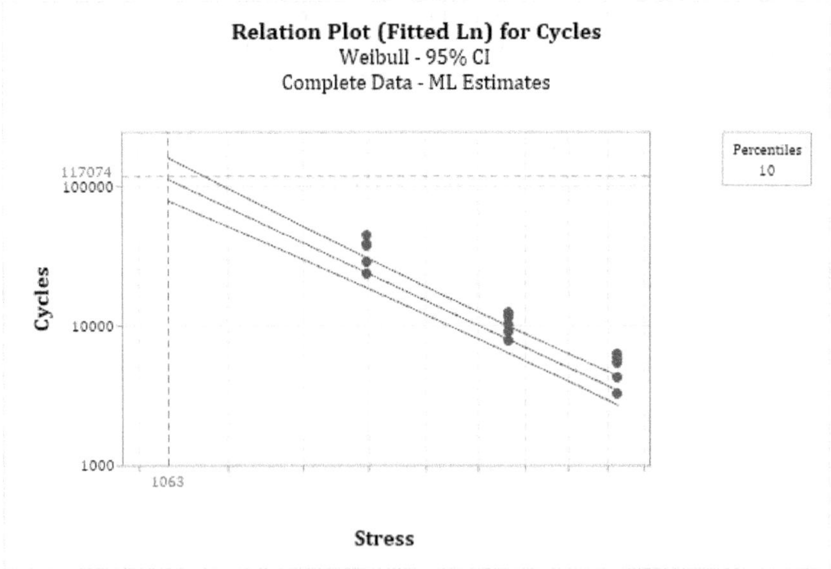

(c)

Figure 4.7. A relation plot for estimating the operating life from experimentally observed accelerating life using inverse power law relationship for **(a)** Specimen A, **(b)** Specimen B, **(c)** Specimen C

4.3.2 Application of Lundberg-Palmgren Model

The empirical model that has widely been approved for contact load components is the inverse power law relationship **[27, 92].** In the current study, the model is applied and tested for OWC for each elevated torque value. Due to the unique nature of contact fatigue and the inability to relate to classical fatigue, most of the prior work in evaluating the lives of rolling bearings was based on empirical investigations and resulting formulas **[44-49].** The fundamental work of Lundberg and Palmgren provided the first theoretical basis for formulating a bearing life model **[30-31]**. The L-P model predicts crack initiations at a particular depth of maximum orthogonal shear stress in the subsurface level at a weak volume point of the material due to contact loads. Such weak points are presumed to be stochastically distributed in the material volume. The Weibull statistical theory **[32-33]** is applied to the stressed volume to obtain the probability of survival of the volume from subsurface-initiated fatigue. The L-P model for the life test can be considered a special characterization of the inverse power model. The material parameters for standard-bearing steel are utilized to develop the load-life relationship. The load life equation for bearing proposed by L-P in terms of the 90% probability of survival of an identical group of bearings **[30]** is given by **Eq. (4.2).**

$$L_{10} = \left(\frac{C}{P}\right)^n \tag{4.2}$$

Where, L_{10} is the life for a 10% probability of failure, P is the equivalent load on bearing contacts, n is the exponent equal to 3 for ball bearings having elliptical contact area, 4 for pure line contacts, and 10/3 for modified line contacts as in the current analysis. C is the basic dynamic capacity given by **Eq. (4.3) [153],**

$$C = \lambda\, M \frac{(1 \mp \gamma)^{29/27}}{(1 \pm \gamma)^{1/4}} \left(\frac{\gamma}{\cos\alpha}\right)^{2/9} d_r{}^{29/27} l^{7/9}\, z^{-1/4} \tag{4.3}$$

$$\text{Where, } \gamma = \frac{d_r \cos\alpha}{d_m},\ d_m \approx D + d_r$$

Here, d_r is roller diameter in mm, l is roller length in mm, z is the number of rollers, α is the clamp angle in degrees, D is the mean diameter of inner race in mm, d_m is the pitch diameter in mm, λ is the L-P reduction factor to account for the stress concentration for edge loading of rollers and non-uniform load distribution along the roller length. The value of λ depends upon the size and geometry of contacts. M is the material constant. Note that C is the constant for a given clutch geometry and material.

The force analysis at contact regions uses Hertz's cylindrical contact theory **[11]**. The values of λ for different sizes of clutch considered in the current study, along with material parameters **[153]**, are listed in **Table 4.9.**

Table 4.9 Material parameters for OWC specimens

Parameter	**Symbol**	**Value**
Modulus of Elasticity	E	200 GPa
UTS of AISI 52100	S_{ut}	2.25 GPa
Reduction factor	λ	0.6-0.8
Material constant	M	552 $N/mm^{50/27}$

The L_{10} life in millions of revolution for roller line contacts by considering these factors can be expressed by **Eq. (4.4),**

$$L_{10} = \frac{z}{2}(1+\gamma)\left(\frac{C}{P}\right)^{\frac{10}{3}} \tag{4.4}$$

According to Hertz contact theory, the normal load F_n on a line contact can be expressed in terms of maximum contact pressure, p_{max} by **Eq. (4.5),**

$$P = F_n = \left[\frac{\pi \ell R_{equ}}{E_{equ}}\right] p_{max}{}^2 \tag{4.5}$$

$$\text{Where, } \frac{1}{R_{equ}} = \frac{2}{d_r} + \frac{2}{D} \text{ and } \frac{1}{E_{equ}} = \frac{1-\vartheta_1{}^2}{E_1} + \frac{1-\vartheta_2{}^2}{E_2}$$

Here, E_1, E_2 are elastic moduli of the roller and race material, respectively, with ϑ_1, ϑ_2 are the respective Poisson's ratios. Thus, the L_{10} life in millions of revolutions for roller line contact can be expressed by **Eq. (4.6)** as,

$$L_{10} = \frac{B}{p_{max}{}^n} \tag{4.6}$$

Where B is the constant representing clutch geometry and material, and n is the power exponent of the load-life relationship.

The geometrical and material constant B and power exponent n for specimen B are derived for the current specifications of the clutch and substituted in **Eq. (4.6).** The load-life equation can be expressed as **Eq. (4.7),**

$$L_{10} = \frac{2749.17}{p_{max}^{6.67}} \quad (4.7)$$

The maximum contact pressure values are calculated using **Eq. (4.5)** as 3.76 GPa, 3.96 GPa and 4.05 GPa for specimens A, B, and C, depending on the torque transmitting capacity of the respective clutch. The experiments are performed for different accelerating conditions to speed up the experiments.

Due to the complex state of stress experienced by roller one-way clutches, the conventional fatigue life equations cannot directly be applied. Fatigue life is sensitive to the applied load; therefore, the nature of loading and the geometry of rolling contacts also influence the life. Hence, it is necessary to investigate the effect of such parameters on the fatigue life of OWC. **Figure 4.8** shows the fatigue life variation plot for accelerating contact pressure values. It indicates that the life decreases with increasing contact pressure due to the increased magnitude of loading.

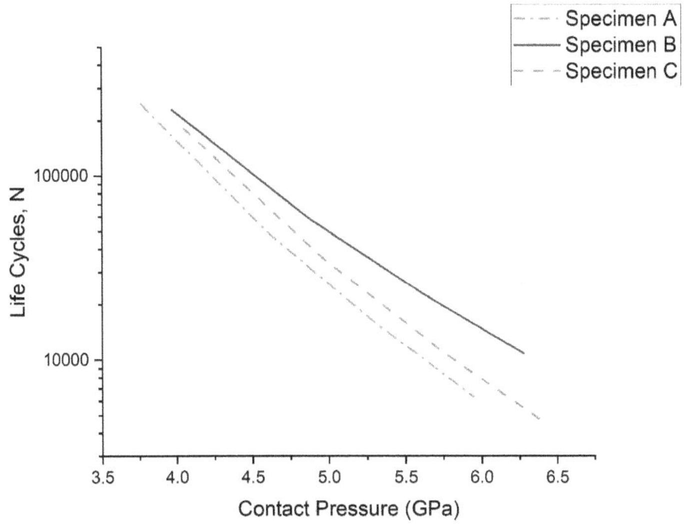

Figure 4.8. Variation of fatigue life with contact pressure.

One of the main objectives of the current study is to define the load-life relationship for OWC applications. The alteration in pressure due to contact force is responsible for the variation in the life of OWC. Thus, the torque transmission capacity directly depends on the amount of contact pressure generated within the permissible limit of the material. The inverse power law relationship is used to predict the life at the operating torque level for OWC applications, and the load-life **Eq. (4.8)** in terms of transmissible torque for specimen B is derived by,

$$L_{10} = \left(\frac{9980.17}{T}\right)^{3.33} \tag{4.8}$$

A similar analysis is carried out for specimens A and C of different sizes, and the corresponding experimental life data is collected. The load-life characteristics are defined using the L-P model displayed in **Figure 4.9** for comparison.

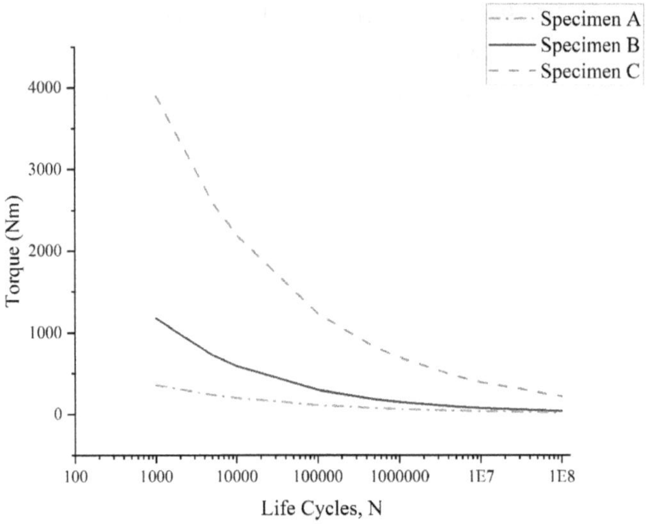

Figure 4.9. Load-life relationship obtained for OWC using the L-P model.

4.3.3 Effect of contact angle variation and surface friction on fatigue life

The basic principle of contact geometry for OWC indicates the importance of clamp angle for inducing normal load responsible for contact pressure variations at the contact surfaces. Varying the geometry of contacts results in varying contact angles with the races that enable the rollers to come into contact at different angular positions. It alters the amount of load transferred to the races and ultimately shifts the contact pressure values. It eventually causes variation in life measures. Thus, fatigue life theoretically relies on the sprag geometry and clamp angle design as presented in the flow diagram as **Figure 4.10**

Figure 4.10 Flow diagram representing the dependency of fatigue life on clamp angle

During torque transmission, through elastic deformations, the position of the roller in the clamp pocket changes, as does the clamp angle. The geometrical principle applied to OWC design observes that the clamp angle value must be defined to guarantee secure gripping from freewheel operation to locking or vice-versa. If the angle falls beyond the clamp angle limit, there is a considerable chance of slippage. **Figure 4.11** shows the upper limit of the clamp angle derived for the specimen under study from corresponding force calculations.

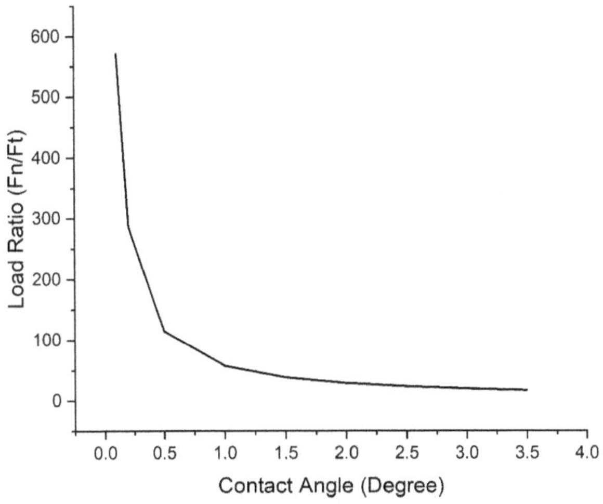

Figure 4.11 Contact force ratio as a function of contact angle.

Practical limits are also imposed on minimum value since the elastic strain energy release of the clutch system tends to jam the rollers and prevent smooth disengagement. Further reduction amplifies the normal load transfer at the contact, which, in turn, increases the amount of hoop stresses in the races and eventually reduces the fatigue life of OWC. The applied torque imposes a higher normal load and high stresses for small angles. Conversely, the larger the strut angles, the smaller the contact stresses, and the clutch becomes more sensitive to the surface finish. Thus, the clamp angle is designed based on the material limits and surface conditions. In the current analysis, the specimens are designed for the minimum clamp angle value of 3.2°. The analysis is extended by increasing the value to the adherence limit, and respective life data are estimated, as shown in **Figure 4.12.**

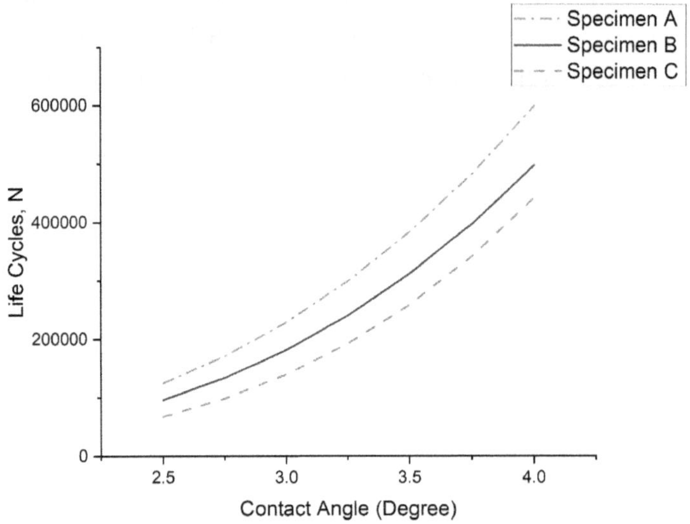

Figure 4.12 Variation of fatigue life with contact angle.

Figure 4.13 (a) presents a Weibull plot of fatigue lives in the presence of different surface friction coefficient values. For the current analysis, the value of friction coefficient μ for inner race surface contact is measured as 0.05 for respective surface roughness. Surface friction varies due to different machining and lubricating conditions within the range of 0.05 to 0.08. The results indicate that the failure probability increases with increasing value of friction. The fatigue life predicted for specimen B from accelerated analysis for different friction values is depicted in **Figure 4.13 (b).** It is observed that the life decreases with the increasing coefficient of friction since the location of subsurface crack initiation moves relatively closer to the surface, and the subsequent time is reduced for the crack propagation to advance to the surface.

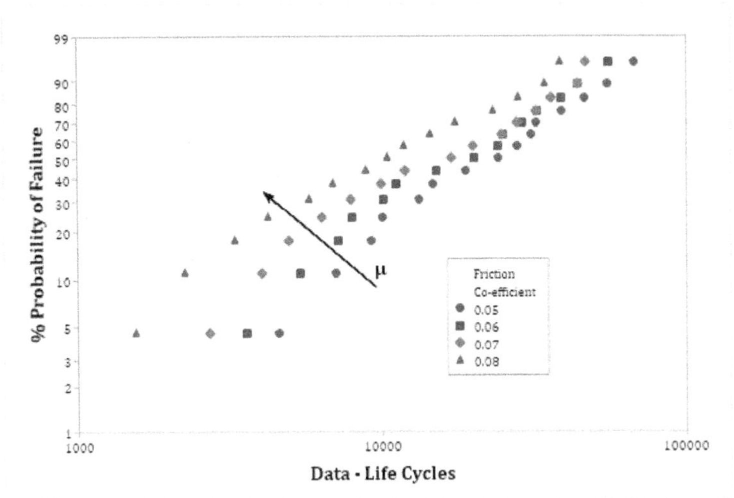

(a)

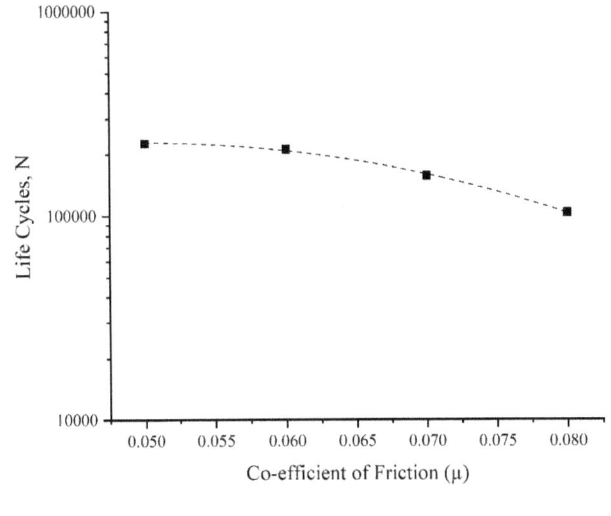

(b)

Figure 4.13 (a) Weibull plot for fatigue lives for different surface conditions **(b)** Effect of varying surface friction for predicted life of OWC

4.4 Experimental Fatigue Life Estimation using ALTA and Life Comparison

The load-life relationship for roller OWC application is derived by performing fatigue experiments on different clutch sizes. In order to achieve this, fatigue life data of roller OWC manufactured from case-hardened alloy steels for each accelerating torque level are recorded and compared with the conventional life prediction model. **Figure 4.14** illustrates the load-life relationship obtained using the L-P theory. Fatigue lives are obtained from accelerated tests using a 2-parameter Weibull estimate for different clutch sizes, considering five samples for each elevated torque level. These results are also incorporated in the same figure for comparison. The fatigue lives are found to fit an inverse power law model with respect to torque levels. The considerable variation in life data from this comparison is due to various reasons. The failure of the OWC is considered when the transmissible torque value on the torque sensor falls below 5% of the specified torque. The L-P model is applied to the available geometrical and material data for the clutch specimens. The load-life relationship developed using the L-P model is based on the empirical relationship expressed by **Eq. (4.2-4.5)**, which does not consider the effect of surface traction, surface irregularities, scratches, dents, manufacturing flaws, and surface-initiated failures **[27]**. Also, there are certain assumptions in the L-P model when developing S-N curves, i.e., reduction factor, material constant, etc.

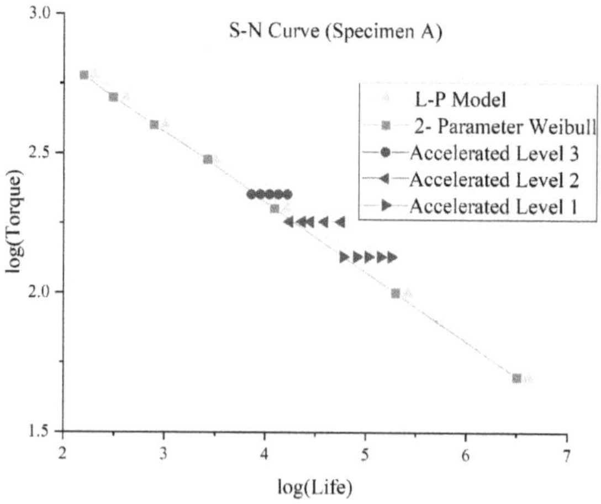

(a)

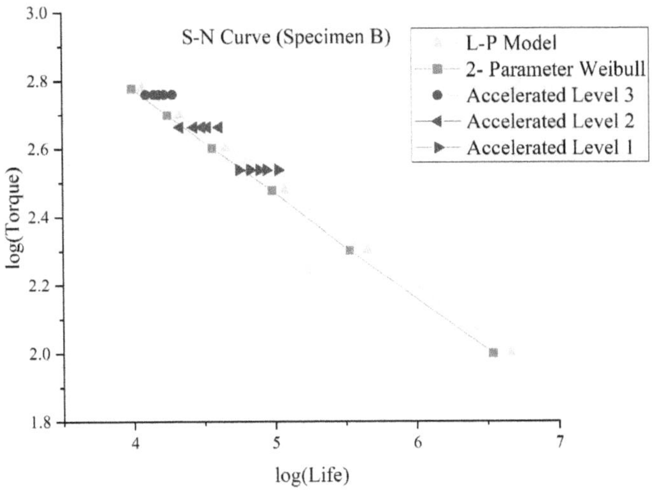

(b)

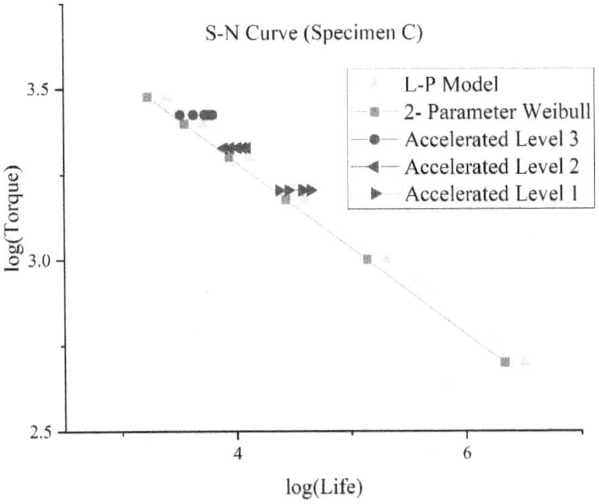

(c)

Figure 4.14. Load-life relationship for **(a)** Specimen A, **(b)** Specimen B, **(c)** Specimen C.

On the other hand, the assumptions made for performing accelerated analysis may cause variation in life results while extrapolating from accelerated results to operational torque value. In addition to this, Weibull analysis is carried out from the experimental data/life cycle counts, which incorporates the effects of actual working conditions of the contact regions, which is not considered directly in the L-P model/Inverse power model in **Eq. (4.6).** Hence, there exist 3-5% life variations have been observed while comparing 2-Parameter Weibull and L-P life models.

4.5 Failure Analysis of Fractured Specimen

Fractography is the study and interpretation of fractured or damaged surfaces, often utilized to understand the nature of the damage and its governing mechanism. Most of the HCF components may fail due to surface or sub-surface-initiated failure mechanisms under repeated loads. In order to investigate the respective failure criterion or the origin of the damage, it becomes necessary to study these components in fractured or damaged conditions. OWC under investigation fails under contact fatigue due to repeated engagements. The failed ramp surfaces of the clutch specimen for the current study are depicted in **Figure 4.15**. The Hitachi model S-3400 N of high-resolution scanning electron microscope, illustrated in **Figure 4.16**, with a resolution of 10 mm at 3 kV, is utilized to observe the fracture surface morphology.

(a) **(b)**

Figure 4.15. Failed clutch specimen **(a)** Inner race ramp surface, **(b)** Outer race surface

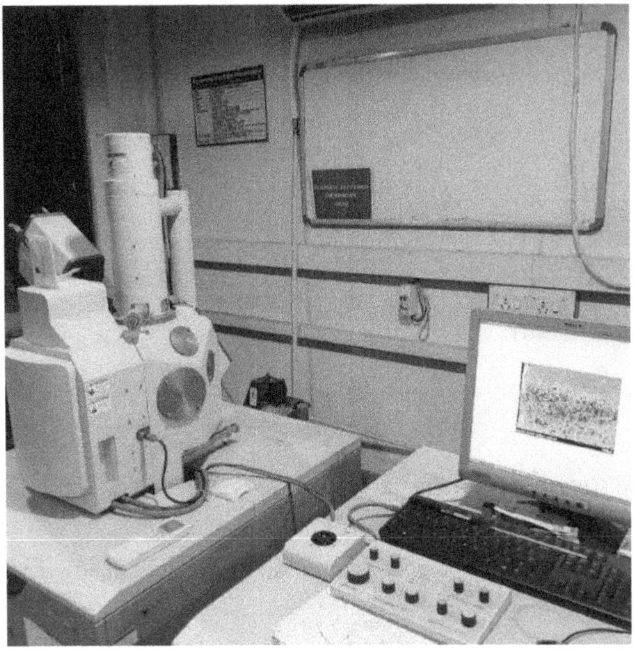

Figure 4.16. Fractography by Scanning Electron Microscope

The failure of OWC by repeated loading and unloading is observed due to the onset of subsurface crack initiation and propagation in addition to scuffing or smearing of the contact surfaces. The excessive wear and fracture of the sliding surface of case-hardened inner race due to repeated engagement and disengagement is therefore considered a failure criterion, which results in a steeper contact angle, enabling OWC to slip during torque transmission. The macro-observation of surface impressions on inner race contact subjected to repeated loading is shown in **Figure 4.17 (a).** The micro dents on the inner race surface due to high contact stress developed at the contact regions generate micro spalls due to a small amount of sliding when rollers are forced into the wedge pocket, as shown in **Figure 4.17 (b).** Further analysis indicates that the clutch material had significant inclusions **[154],** which serve as the stress risers and increase the propagation rate of damage in the material domain.

(a) **(b)**

Figure 4.17. Smearing/scuffing of the inner race with **(a)** Macro-profile **(b)** SEM micro-profile

The fractured surface appeared in the case, and core zones are examined in an SEM, as shown in **Figure 4.18**. The case-carburized surface characterizes inter-granular morphology due to higher carbon diffusion, and core fracture displays dimple morphology due to low carbon content, which is observed from SEM micrographs illustrated in **Figures 4.18 (a)** and **4.18 (b),** respectively. During torque transmission from one race to another, rollers are initially forced into the races with the help of springs, which facilitates them partially sliding before complete engagement to transmit the required amount of torque. Sliding occurs when the lubricant breaks down and transfers the load. Such sliding can considerably alter the stress distribution and the location of maximum stress moving closer to the surface **[155].**

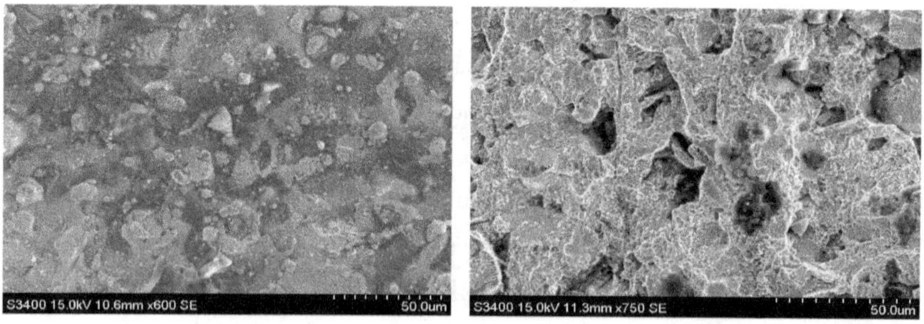

(a) **(b)**

Figure 4.18. SEM micrographs of the inner race for **(a)** case-carburized surface region and **(b)** core region

The sub-surface cracks, therefore, initiate and branch out further and produce micro-pits due to the fracture of sliding surfaces, as shown in **Figure 4.19**. In addition, the surface irregularities and foreign particles on the surface together can act as additional stress risers to develop surface cracks, ultimately leading to localized surface damage. The fractography analysis shows micro cracks initiated at the surface due to the surface traction effect, transient sliding of contact surfaces, and other micro-surface irregularities. Surface friction plays a significant role in localized surface crack initiation and propagation **[156]**. These cracks ultimately form micro flakes due to progressive damage from repeated loading and surface pitting, as shown in **Figure 4.19 (a)**. **Figure 4.19 (b)** illustrates the layer of surface broken out due to repeated engagement under the effect of adhesive wear, which keeps the stress concentration rising in the vicinity of the contact region. The material is eventually liberated in the form of micro flakes, which result in sharp patterns appearing on the surface in the sliding direction. Such progressive material degradation due to adhesive wear manifests itself as one of the prime reasons for surface depression **[157-158]** or a shallow wear scar on the inner race contact regions, ultimately resulting in a steeper contact angle. Such a steeper contact angle requires a larger friction coefficient to prevent slip. Ultimately, the contact angle becomes so large that the available friction cannot transmit the required torque, and the clutch eventually fails. The deformation bands also appear after several load cycles due to the simultaneous effect of loading sliding.

(a) **(b)**

Figure 4.19. Inner race failure observed through SEM due to **(a)** Micro cracks propagated to the surface with repeated surface pitting resulting in surface dents and **(b)** progressive material removal in the form of micro flakes

4.6 Summary

This chapter focused on a series of experiments performed on OWC specimens. The test plan for conducting experiments and the relevant test methodology are presented to evaluate the fatigue life. The experimental fatigue data are studied using accelerated life test analysis to predict the life at the available loading and operating conditions. The conventional Lundberg-Palmgren fatigue model is employed on the available clutch geometries to estimate the fatigue life and compared with the accelerated model observed through experiments. The effect of various parameters such as contact pressure, contact geometries and surface friction is investigated. At last, the failed clutch geometries are examined using a scanning electron microscope to understand the governing failure mechanisms of the clutch. It is found that the fatigue life model developed through accelerated tests exhibits an inverse power load-life relationship with respect to torque levels and is similar in structure to the original L-P model. In addition, an accelerated model developed is found to follow Weibull distribution as a life characteristic at elevated torque levels. The failure examination reveals the typical forms of wear observed in OWC, like surface pitting, smearing, and scuffing, which are responsible for clutch failure.

Chapter 5#

Numerical Analysis for Fatigue Life Estimation using Continuum Damage Mechanics based Finite Element Model

This chapter presents a numerical strategy to estimate the fatigue life of OWC. A damage-coupled FE analysis for the OWC contact region is carried out to estimate the crack initiation and propagation lives. Case hardening and residual stress effects on fatigue life are also discussed. The load-life relationship developed using a numerical model is compared with the experimental results.

GRAPHICAL ABSTRACT

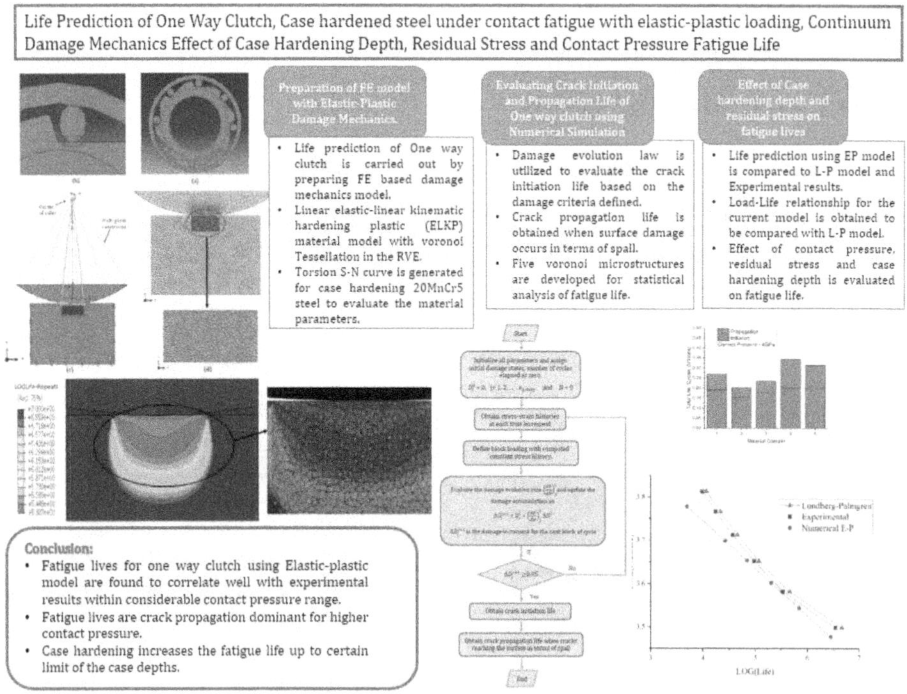

#*Karan A. Dutt*, *Shashikant J. Joshi, Dhaval B. Shah, Shashikant B. Soni, Deepak Prajapati (2024). "Life Prediction of One Way Clutch for Case Hardened Steel Under Contact Fatigue with Elastic-Plastic Loading Using Continuum Damage Mechanics." Iranian Journal of Science and Technology: Transactions of Mechanical Engineering.*

5.1 Numerical Modeling Approach

As discussed in Chapter 2, sub-surface initiated failure for bearing steel or case hardening steel is influenced by material microstructure. It includes a heterogeneous mixture of carbide, austenite and martensite, along with non-metallic inclusions. Such inclusions act as local stress risers and serve as crack initiation regions. Over the years, advancements in manufacturing technologies have improved material qualities with reduced presence of these inclusions. Even if such inclusions have been removed 100% from the material, weak planes in the material are randomly oriented and distributed throughout the material, which is prone to generate random cracks and eventually cause material failure under fatigue. This topological disorder in the microstructure model is stochastic in nature and has a significant influence on crack initiation and propagation lives of contact fatigue. This chapter describes the numerical approach and FE simulation strategy for estimating the fatigue life of material under investigation using a damage mechanics-based microstructure model.

5.1.1 Microstructure topology

The scatter in fatigue lives due to microstructure randomness in material volume is taken under consideration by generating Voronoi microstructure topology in the material model under study. A Voronoi tessellation technique randomly generates a distinct number of points.

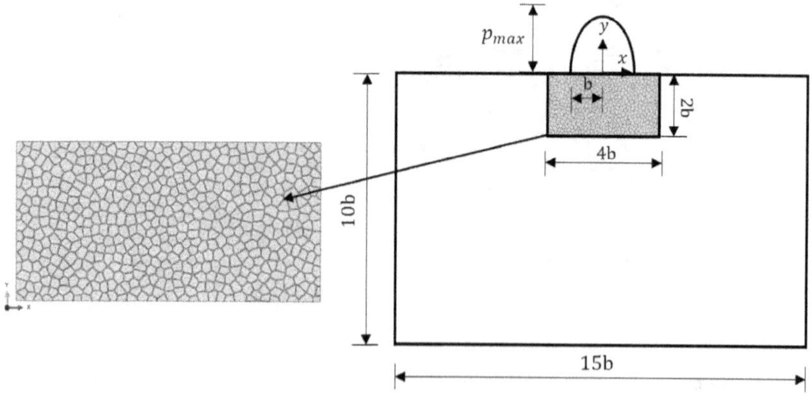

Figure 5.1. Representative Volume Element (RVE) with Voronoi microstructure topology.

A set of boundaries is created around these points closer to the point than other regions, which ultimately results in figures of polygons known as Voronoi cells. A representative volume element (RVE) with Voronoi cells is shown in **Figure 5.1**, where RVE is defined as the critically stressed region for the one-load cycle. The microstructure randomness of the material domain is considered in RVE using the Voronoi tessellation technique. This region is restricted to RVE using tie constraints to reduce computational expenses. The RVE dimensions are governed by the type of Hertzian contact loading, contact geometries and available traction **[159].** In the current investigation of line contact, the RVE dimensions shown in **Table 5.1** are considered as a function of half contact width 'b' since subsurface stress gradients are negligible beyond these values **[75].** The dimensions of the simulation domain are determined heuristically to reduce computation time and do not have any significant effects on result prediction. Five distinct Voronoi simulation domains are generated and taken into consideration for analysis.

Table 5.1. FE-RVE dimensions

Dimensions	Microstructure domain	FEA domain
X	(-2b to 2b)	(-7.5b to 7.5b)
Y	(0 to -2b)	(0 to -10b)

5.1.2 FE-based contact model

A 2-dimensional FE model of a one-way roller clutch is prepared in the software package ABAQUS. The FE model is represented by Hertzian loading and further extended to elastic-plastic loading on a semi-infinite half space. The Hertzian pressure distribution at the contact region is applied as a non-uniform surface traction vector defined by **Eq. (5.1) [11],**

$$F_y(x) = p_{max}\sqrt{1-\left(\frac{x-x_c}{b}\right)^2} \tag{5.1}$$

Where $F_y(x)$ is the normal component of contact traction, p_{max} is maximum pressure at the centre, (x_c,y_c) is the coordinate of the centre at maximum pressure, and 'b' is the half-contact width.

The surface friction is defined in the simulation domain by specifying contact conditions with co-efficient of friction μs by **Eq. (5.2),**

$$F_x(x) = \mu_s \left| F_y(x) \right| \tag{5.2}$$

The roller is 7 mm in diameter, a portion of which is extracted along with a section of inner race from the entire OWC assembly to model the contact and allow reasonable computational time, as shown in **Figure 5.2**. For loading uniformity in the axial direction, the FE model is simplified into line contact and the analysis is carried out for plane strain condition. The centre of the roller is assigned a reference point. The reference point is then coupled to roller geometry using Multi-Point Constraint (MPC) to apply loading and boundary conditions. The lower boundaries of the model are kept fixed by applying zero displacements. Periodic boundary conditions are applied to the right and left surfaces for identical displacements **[66]**. The contact regions at the roller and raceway are finely meshed with plane strain CPE4R type finite elements with a minimum mesh size of 0.02 mm, while the regions other than the contact are coarsely meshed.

The contact simulation in the current study is executed in two steps, applying the normal load with partial sliding between the contacts as,

i. The roller is placed at the centre of the inner race RVE, and displacement along the y-axis is allowed at the MPC point when applying contact load.
ii. A rotational displacement of 0 to 0.05 radians amplitude is applied at the MPC point to provide sliding movement.

The contact stress analysis is performed for one loading cycle to get the stress histories. **Figure 5.3** shows the distribution of stress components for a point corresponding to the maximum orthogonal stress depth in a single load cycle. It reveals that the stress components are non-proportional. It means that the peaks of normal and shear stresses do not coincide. Also, the complete reversal of shear stress is observed, and the normal stresses become compressive.

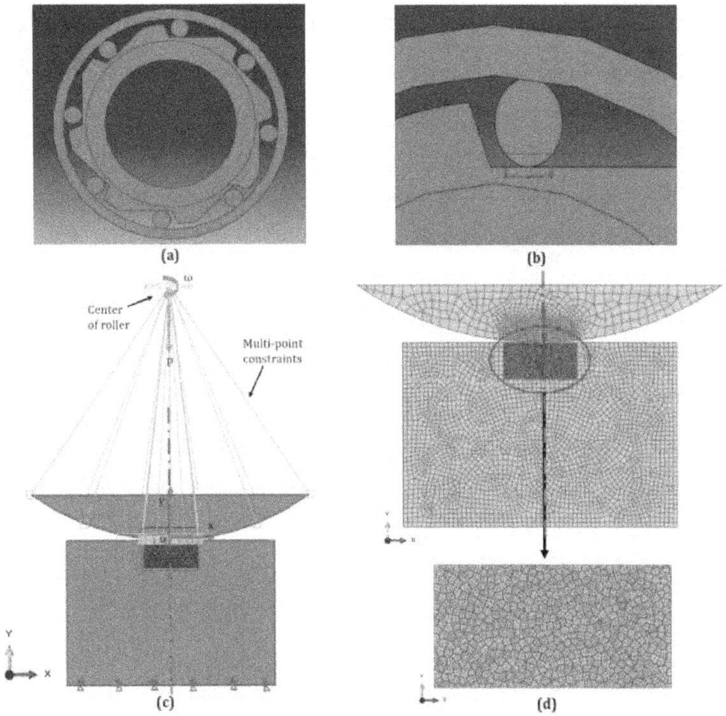

Figure 5.2. FE modelling **(a)** one-way clutch assembly **(b)** RVE under study **(c)** MPC constraints with boundary conditions **(d)** Meshing

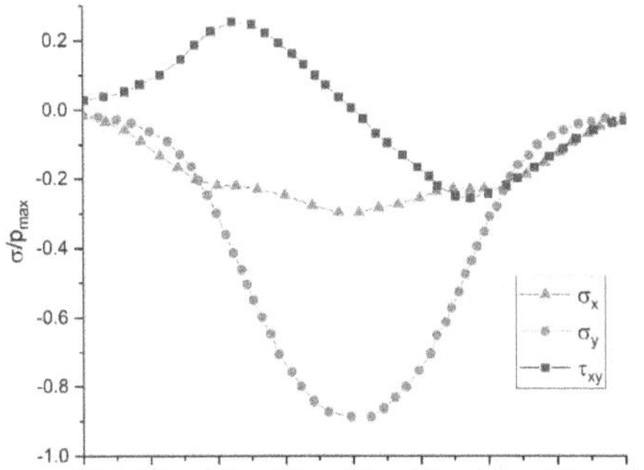

Figure 5.3. Stress distribution for a subsurface point located near maximum orthogonal shear stress

5.2 Continuum Damage Mechanics-based Material Model

Damage mechanics involves the investigation of micro-mechanisms from a continuum mechanics point of view represented by an RVE, as shown in Figure 5.4 [53], which is responsible for material failure in an empirical fashion. The development of distributed cavities in the material model by damage and fracture can be investigated in the framework of continuum damage mechanics (CDM) by the following procedure.

1. Representing the corresponding damage state using a damage variable D(x).
2. Formulating the governing equation (i.e. evolution equation) to develop the damage variable.
3. Formulating an equation displaying the mechanical behaviour of the damaged material (i.e. constitutive equation).
4. Solving the initial-value and boundary-value problem by using these equations.

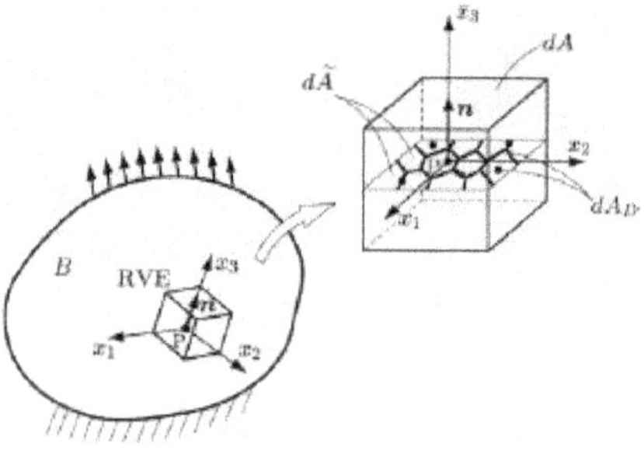

Figure 5.4. Representative Volume Element (RVE) under CDM **[53]**

5.2.1 Modelling of case carburization steel

A set of experiments to characterize the material properties of case hardening steels have been performed by Branch et al. **[160]**. The relationship of hardness with yield strength and ultimate tensile strength was investigated by Cahoon et al. **[161]**. Brief research on compiling the hardness with strength for various steel microstructures was executed by Pavlina et al. **[162]**. Using the least square regression technique, yield strength (S_y) (MPa) can be evaluated as a function of Vickers hardness (H_v) (kgf/mm^2) **by Eq. (5.3),**

$$S_y = -90.7 + (2.876 * H_v) \tag{5.3}$$

The mechanical stress-strain histories of the finite element model subjected to line contact are correctly captured by proposing an appropriate material model. The highly localized compressive stress in roller OWC often produces micro-plastic deformation at the contact region. In addition to this, the hardening process also alters the material's plasticity. So, an elastic-plastic material model must be incorporated into the analysis to consider the hardening effect at contact regions. The type of hardening rule under cyclic load can be both isotropic and kinematic and generally depends on the material properties. The isotropic hardening rule accurately describes the behaviour of ductile materials subjected to large plastic deformations where the yield stress continuously increases with each load cycle. The resistance to these deformations also increases consistently. The yield function f(σ, k) for isotropic hardening depends upon the state of stress σ and isotropic hardening variable k. In the current analysis, for the case-hardening steel, the contact fatigue nearly exhibits the quasi-brittle behaviour **[163]** in the absence of isotropic hardening, for which the use of the kinematic hardening rule is more prominent.

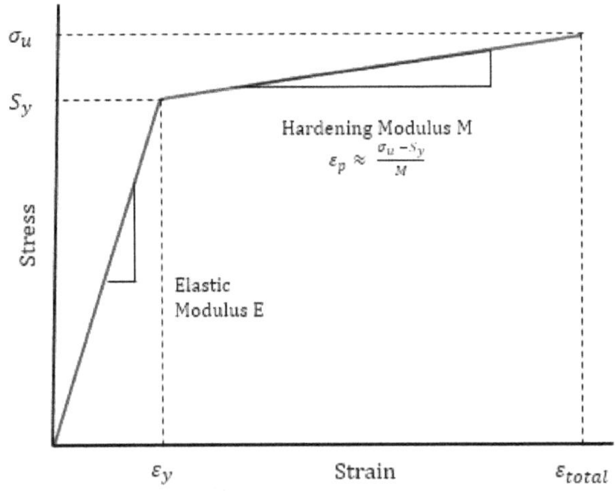

Figure 5.5 ELKP material model representation

Hahn et al. **[164]** proposed that a power law fit of any elastic-plastic stress-strain response can be approximated and simplified into a bilinear representation of a linear elastic-linear kinematic hardening-plastic (ELKP) material model. The analysis becomes more straightforward with bilinear representation by reducing the number of parameters in such a model. The variation in hardness and residual stress can also be emphasized easily with the ELKP model. In the current study, the stress-strain behaviour of case hardening steel is accurately modelled using a similar approximation by the UMAT subroutine to capture elastic-plastic material behaviour. **Figure 5.5** presents the corresponding bilinear representation of the ELKP model.

Fatigue due to repeated loading on contact surfaces results in material degradation through crack initiation and propagation. The contact surfaces also suffer from smearing, i.e., scuffing out the weak layers due to transient sliding of the rollers under load. Such material degradation is taken into account by extending the current model in the damage mechanics domain. The damage mechanics has been proved to be a suitable approach to account for such failure mechanisms considering a damage variable D and the corresponding relationship takes the following form as constitutive equation **Eq. (5.4(a)) [53]**,

$$\sigma_{ij} = C_{ijkl}(I_{klnm} - D_{klnm})\,\varepsilon_{mn} \tag{5.4(a)}$$

Where σ_{ij} is the stress tensor, C_{ijkl} is the material stiffness tensor containing the elastic constants, I_{klnm} identity tensor, D_{klnm} damage tensor and ε_{mn} is the strain tensor.

Here, any damage initiation leads to respective stiffness degradation of the material. The undamaged condition of the material can be represented by damage parameter D = 0, and the fully damaged condition indicates micro crack initiation at D = 1.

For the isotropic damage state, the damage tensor reduces to a scalar and simplifies as **Eq. (5.4(b))**

$$\sigma_{ij} = C_{ijkl}(1 - \mathrm{D})\,\varepsilon_{kl}\,,\; 0 \leq \mathrm{D} \leq 1 \tag{5.4(b)}$$

So, the presence of damage leads to a respective stiffness reduction for the material under consideration. The evolution of damage in a high cycle fatigue can be related to the stress level at the concerned material point through the non-linear differential **Eq. (5.5(a))** in the form as,

$$\frac{dD}{dN} = f(\sigma, \mathrm{D}) \tag{5.5(a)}$$

Where σ is the stress measure, D is the damage variable, and N is the number of stress cycles. All the above equations follow the concept of effective stress, which is the state of stress after each material degradation due to fatigue load and the damage mechanics principle of strain equivalence. The effective stress induced on the undamaged area satisfies the following constitutive relationship in terms of damage state **Eq. (5.5(b)) [53]**,

$$\tilde{\sigma}_{ij} = \frac{\sigma_{ij}}{(1-D)} == C_{ijkl}\ \varepsilon_{kl} \tag{5.5(b)}$$

Similarly, the principles of damage mechanics are applied to the elastic-plastic material model as a coupled solution **[60-61]**. By considering the Mises yield condition with kinematic hardening (back stress variable) and incorporating the damage state, the relevant equivalent stress σ_{EQ} and the damage yield function f can be written as **Eq. (5.6(a-d))**,

$$\sigma_{EQ} = \sqrt{\frac{3}{2}\left(\sigma_{ij}^{D}\sigma_{ij}^{D}\right)}, \tag{5.6(a)}$$

$$f = (\tilde{\sigma} - \alpha)_{EQ} - \mathrm{R} - \sigma_Y \tag{5.6(b)}$$

Since in the absence of isometric hardening, the value of R is taken as zero.

$$f = (\tilde{\sigma} - \alpha)_{EQ} - \sigma_Y \tag{5.6(c)}$$

$$(\tilde{\sigma} - \alpha)_{EQ} = \sqrt{\frac{3}{2}\left(\frac{\sigma_{ij}^{D}}{1-D} - \alpha_{ij}\right)\left(\frac{\sigma_{ij}^{D}}{1-D} - \alpha_{ij}\right)} \tag{5.6(d)}$$

Where α is the back stress tensor, which represents the centre of the yield surface, $\tilde{\sigma}$ is the effective stress tensor, $\tilde{\sigma}_{ij}^{D}$ is the components of the deviatoric tensors of stress $\tilde{\sigma}$, which are respectively expressed by **Eq. (5.6(e))**,

$$\alpha_{ij}^{D} = \alpha_{ij} - \frac{1}{3}A_{kk}\delta_{ij},\ \tilde{\sigma} = \frac{\sigma}{1-D} \text{ and } \tilde{\sigma}_{ij}^{D} = \tilde{\sigma}_{ij} - \frac{1}{3}\tilde{\sigma}_{kk}\delta_{ij} \tag{5.6(e)}$$

It is noted that the shear stress component experiences the stress reversal in cyclic load and is responsible for the failure of OWC in fatigue. The damage evolution at the material point depends on the material model and is governed by the law based on the failure mechanism. For the elastic model, damage depends on the state of stress, whereas the elastic-plastic model depends on the state of stress and accumulated plastic stain.

For the elastic case, by considering maximum shear stress reversals ($\Delta\tau_{critical}$), the contact fatigue in case hardening steel nearly exhibits the quasi-brittle behaviour **[164]**, for which the damage evolution law is derived by **Eq. (5.7)** as,

$$\frac{dD}{dN} = \left[\frac{\Delta\tau_{critical}}{\tau_{r0}\left(1-\frac{\sigma_m}{S_{us}}\right)(1-D)}\right]^m \tag{5.7}$$

Where, σ_m is the mean stress and S_{us} is the ultimate shear strength, τ_{r0} is material fatigue resistance, m is the material parameter. The mean stress variation due to residual stress is considered in damage evolution law for modelling the contact fatigue derived by Shen et al. **[165]**.

The damage evolution for plastic case **[53]** by considering maximum Von Mises stress ($\sigma_{V\,Max}$) and accumulated plastic strain ($\dot{p}$) is given by **Eq. (5.8)**,

$$\frac{dD}{dN} = \left[\frac{{\sigma_{V\,max}}^2}{2ES_0\left(1-\frac{\sigma_m}{S_{us}}\right)^2(1-D)^2}\right]^q \dot{p} \tag{5.8}$$

Where S_0 and q are the material parameters characterized by their resistance to fatigue damage, and E is the modulus of elasticity.

5.2.2 Identification of material parameters

During torque transmission from one race to another, the contact pressure is generated at contact regions. In order to improve the durability of the material under such heavy loads, the race materials require surface treatments. Therefore, the OWC races are manufactured from case-hardened steel using a carburization process to improve fatigue and wear resistance. Table 1.3 illustrates the chemical compositions required for inner race material and the values measured using an optical emission spectrometer (OES).

The damage evolution laws expressed by **Eq. (5.7-5.8)** for elastic and elastic-plastic models, respectively, contain two material parameters in each equation. These parameters are required to be extracted from cyclic torsion stress-life data. The reason behind selecting torsion fatigue data for evaluating the parameters corresponding to fatigue life is logical because the damage accumulation mechanism for contact fatigue is shear stress driven **[166]**, equivalent to torsion fatigue. Moreover, the case carburization process at the contact surface increases the carbon content from ~0.2 to ~0.8 per cent of weights up to the required case depth. Also, it is well hypothesized that life standards pertain to bearings made of through-hardened steel with a minimum Rockwell hardness of 58-62 HRC, commonly applicable to case-hardened steels **[167]**. The relationship of hardness with yield strength, ultimate tensile strength, strength co-efficient 'K' and strain hardening exponent 'n' is expressed by **Eq. (5.9-5.12) [161]**. Based on that, the mechanical and case-hardened properties of DIN 20MnCr5 are presented in **Table 5.2 [167-168]**

$$S_y = 0.0012(\text{HB})^2 + 3.3(\text{HB}) \tag{5.9}$$

$$S_{ut} = -99.8 + 3.734(\text{HV}) \tag{5.10}$$

$$\text{n} = -0.271 \ln\left(0.88 - \frac{27.75}{HV}\right) \tag{5.11}$$

$$\text{K} = \frac{HV}{3(0.08)^n} \tag{5.12}$$

Table 5.2 Case-hardened 20MnCr5 steel surface properties **[167-168]**

Parameters	**Values**
Density $\boldsymbol{\rho}$, kg/m^3	7850
Modulus of Elasticity E, GPa	210
Poisson ratio, $\boldsymbol{\vartheta}$	0.3
Tensile strength $\boldsymbol{S_{ut}}$, MPa	2686
Yield strength $\boldsymbol{S_y}$, MPa	2054.8
Strength co-efficient K, MPa	2741.2
Strain hardening exponent, n	0.0462
Rockwell hardness at the surface, HRC	60 ± 2
Rockwell hardness at the core, HRC	40 ± 2
Fracture toughness K_{IC}, MPa.$\sqrt{\boldsymbol{m}}$	80 (Core) 22 (Surface)
% R.A (Residual Austenite)	6.3

The cyclic torsional fatigue parameters for 20MnCr5 steel are evaluated using hardness-dependent uniaxial fatigue properties derived by Kim et al. **[169]** and Josip et al. **[170]**. McClaflin et al. **[171]** showed that shear is the controlling mechanism for torsion fatigue, and crack orientations are mainly observed on the maximum principal plane.

The crack initiation observed is shear-driven, and crack propagation is negligible because of the high hardness. In order to investigate torsion fatigue from available axial data, the best practice is to correlate the axial fatigue data with torsion using the maximum principal strain criterion **[172]** in conjunction with the Roessle-Fatemi model. Roessle et al. **[173]** have evaluated the axial strain life data of uniform hardness specimen derived as the function of Brinell Hardness (HB) by **Eq. (5.13-5.14),**

$$\sigma'_f = 4.25\,(HB) + 225 \tag{5.13}$$

$$\varepsilon'_f = \frac{[0.32(HB)^2 - 487\,(HB) + 191000]}{E} \tag{5.14}$$

Where σ'_f is the fatigue stress, ε'_f is the fatigue strain, and E is the modulus of elasticity.

In the current analysis, the shear fatigue life curve for case hardening steel is obtained by **Eq. (5.13-5.14)** using the maximum principal strain criterion in the absence of fatigue data. **Figure 5.6** shows an analytically generated torsional S-N diagram for case-hardened 20MnCr5 steel. By employing the power law relationship to these data, the fatigue strength coefficient 'τ'_f' and fatigue strength exponent 'a' **[167]** can be evaluated as 2.1379 GPa and -0.1013, respectively, as per Basquin's law **Eq. (5.15),**

$$\Delta\tau = \tau'_f N_f^{\,a} \tag{5.15}$$

Where $\Delta\tau$ is the shear stress reversals, N_f is the number of life cycles to failure. By integrating **Eq. (5.7)** and for equivalency with **Eq. (5.15),** the elastic damage parameters m and τ_{r0} can be obtained in terms of Basquin's parameters,

$$m = -\frac{1}{a} \text{ and } \tau_{r0} = 2\tau_f'\left(1 - \frac{1}{a}\right)^{-a} \tag{5.16}$$

Similarly, damage parameters S_0 and q for the plastic case can be derived by integrating **Eq. (5.8)** and equating to **Eq. (5.15),**

$$\sqrt{3}\,(\tau_f' N_f^{\,a}) = \frac{(2ES_0)^{\frac{1}{2}}}{(2(1+2q)\Delta\varepsilon_p)^{\frac{1}{2q}}} N_f^{-1/2q} \tag{5.17}$$

Here, the range of plastic strain $\Delta\varepsilon_p$ can be represented in terms of ultimate stress (σ_{max}) and yield strength (S_y) by hardening modulus M considering ELKP by,

$$\Delta\varepsilon_p = 2\left(\frac{\sigma_{max} - S_y}{M}\right) \tag{5.18}$$

The plastic strain range is evaluated based on the Coffin-Manson strain-life relationship combined with the Roessle-Fatemi model for elastic-plastic condition given by **Eq. (5.19)**

$$\frac{\Delta\varepsilon_p}{2} = \varepsilon'_f (2N_f)^c \tag{5.19}$$

The plastic strain observed during low cycle fatigue (N_f) values from the portion of the S-N curve as N_1 = 10 cycles and N_2= 1000 cycles are calculated and substituted respectively in **Eq. (5.17)** for solving the parameters S_0 and q.

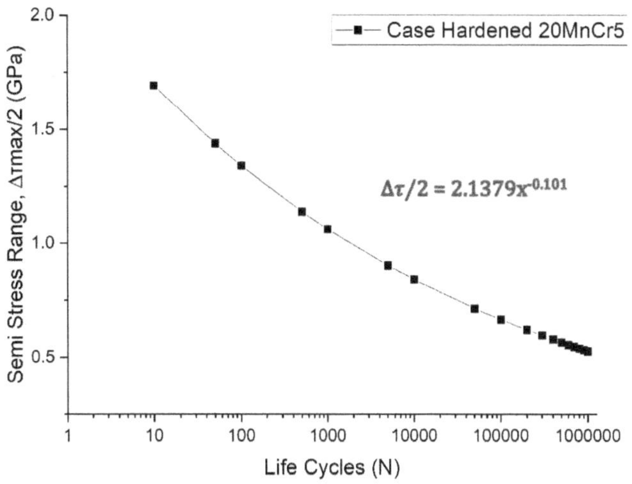

Figure 5.6 Cyclic torsional S-N curve for case hardening steel DIN 20MnCr5

The case hardening material with surface hardness decaying with depth from the surface to core also indicates the respective yield strength variation; therefore, the S-N behaviour of the model is also modified by the hardening process **[174].** In the current analysis, it is estimated that the fatigue strength co-efficient 'τ_f'' linearly varies with hardness and therefore with yield strength without affecting the slope of the curve. Using this assumption, **Figure 5.7** compares S-N curves for varying yield strengths from 2.05GPa to 1GPa with the original yield strength curve (2.05 GPa).

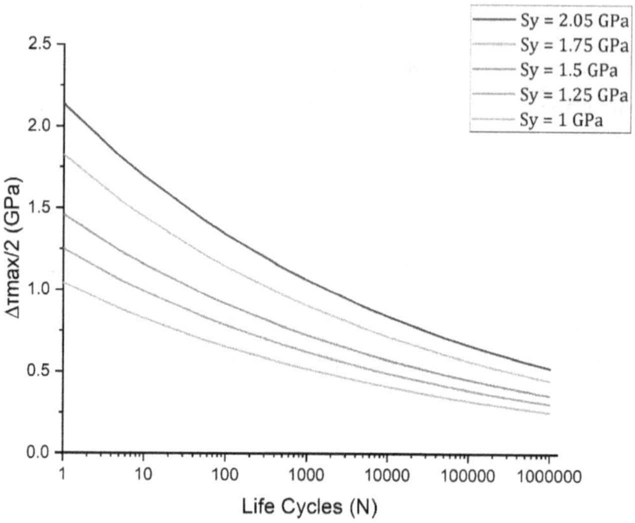

Figure 5.7 Torsional S-N curves analytically produced for case hardening steel 20MnCr5 corresponding to different yield strengths incorporating case depths

5.3 Numerical Implementation for Fatigue Life Prediction

Crack initiation analysis involves the calculations of (i) stresses responsible for releasing the surface energy into crack propagation under repeated loads, (ii) element stiffness degradation, and (iii) computations of progressive damage until the crack initiation criteria, i.e. D = 0.95, is satisfied. Based on the E-P damage model presented in the previous section, a numerical algorithm is employed for characterizing crack initiation using stress-strain histories and damage evolution law. The numerical strategy includes the simultaneous solution of damage-coupled stress fields. The stress-strain histories and equivalent plastic strain for one load pass are obtained from the FE simulation output database and implemented as a MATLAB function solving for damage evolution using Euler's backward method. However, evaluating stress-strain response by the FE model using such an iterative process for each load pass cannot be considered due to HCF. Therefore, "a jump in cycles" method **[53]** is utilized. This strategy considers the state of stress, damage progress, and accumulated plastic strain response for each element constant in each cycle in a defined block of ΔN^i for the ith

block. This way, the damage evolution is represented as piecewise linear with respect to the block of cycles, as depicted in **Figure 5.8.**

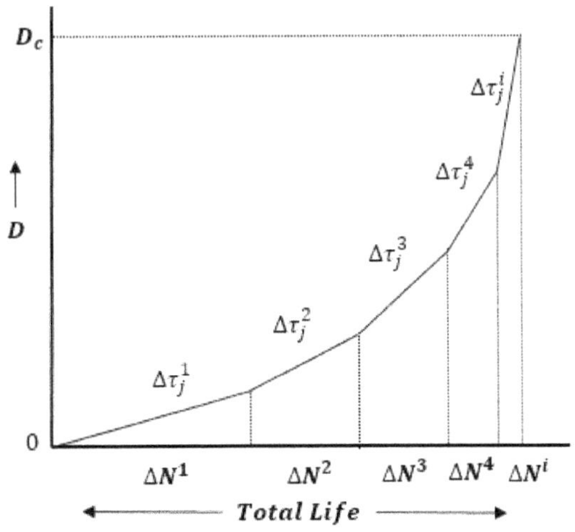

Figure 5.8 Piecewise constant amplitude periodic load with a block of cycles

The damage values are evaluated using a constitutive relationship from the last block of cycles, and the stress-strain field is updated for the next block of cycles. A damage increment value ΔD is selected over the number of cycles such that the simulation can be executed under reasonable computational time (=0.2 for current analysis). The number of cycles in the present block is evaluated by dividing ΔD by the maximum damage rate. The detailed algorithm utilized in the current investigation is illustrated in **Figure 5.9.**

In the present analysis, the damage evolution law defined by **Eq. (5.7)** and **Eq. (5.8)** is concurrently evaluated, and the larger of the two is utilized to modify the damage state. It is noted that the mean stress variation due to residual stress values and the yield strength values are assigned to each finite element as per the case of carburizing depth. This results in a variation of parameters 'τ_{r_0}'and 'S_0'.

Figure 5.10 depicts the damage parameter variations 'S_0' and 'τ_{r0}' with respect to yield strength because of their dependence on fatigue strength coefficient 'τ_f'''. The exponent 'm' and 'q' of the damage rate law for elastic and elastic-plastic models are constants and depend only on the fatigue strength exponent 'a'.

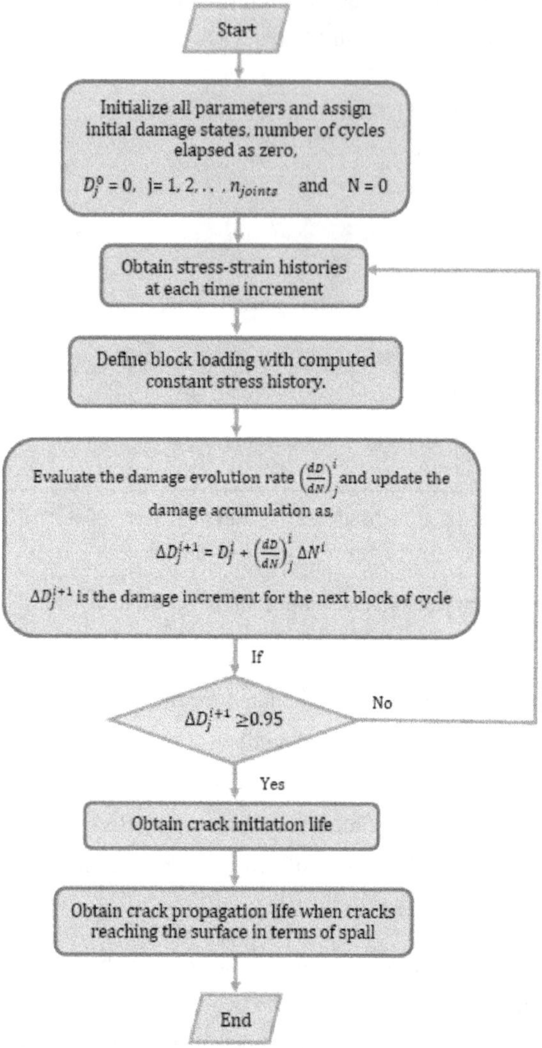

Figure 5.9 Numerical Algorithm for crack initiation

After the crack initiates, the damage evolution rate continues to be applied for the current damage state until the critical damage criterion is reached. The simulation is continued, and thereby, the number of elements that have failed also increases. An FE fatigue tool is utilized to simulate these data to represent all the failed elements for displaying a path, which is considered a path of crack propagation. The number of cycles recorded from crack initiation to surface damage development is considered crack propagation life. The total number of cycles for crack initiation to crack propagation is considered the fatigue life of OWC.

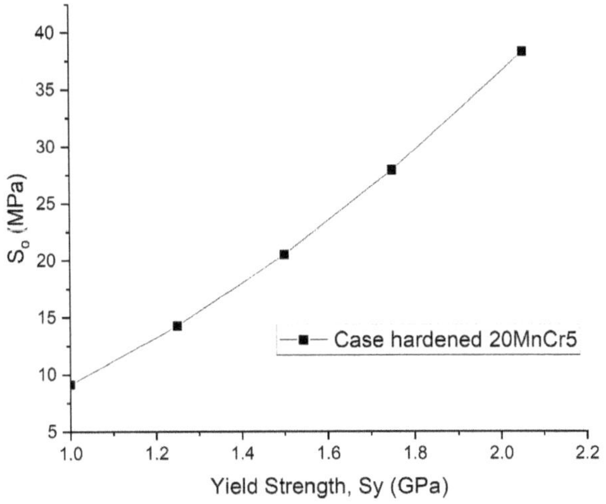

(a)

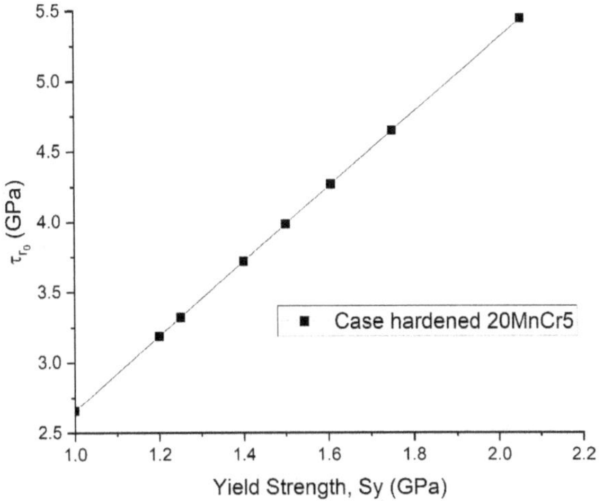

(b)

Figure 5.10 (a) Damage parameter: S_0 variations with a yield strength **(b)** Damage parameter: τ_{r0} variations with yield strength.

5.4 Results and Discussion

During fatigue, material undergoes a localized and progressive damage at microscopic and macroscopic levels which result in stress-strain variations in the material volume. The physics behind the fatigue involves several complex mechanisms i.e. subsurface crack initiation, crack propagation, fracture, that basically depend on the material properties, geometries and the loading conditions. The loading and surface conditions also contribute significant details how the material behaves under cyclic loads. Understanding the material response under varying load condition provides useful insights how material behaves when subjected to non-uniform stress, which is more representative of real world application. The same can also be helpful to estimate the failure modes, crack growth behaviour, damage accumulation and the fatigue life variations.

Moreover, the case carburization process at the contact geometries enriches the surface of the material with carbon to improve the surface properties. This process significantly affects the material behaviour under fatigue subjected to cyclic loads and introduce the residual stress inside the material domain. The residual stress distribution varies based on the case depth. Case depth refers to the thickness of the hardened layer produced by the carburization process. The carbon content shall be maximum at the case region and reduced to the core region up to a particular case depth. It results in maximum hardness at the case and reduced to the core region with a hardness gradient. The hardness gradient also introduce the yield stress variation within the case depth. Thus, the case carburization of the material also plays a critical role in determining the fatigue life under cyclic loads since it affects the surface hardness, residual stress distribution and yield stress variations. A corresponding elastic-plastic FE material model is prepared in ABAQUS considering these two conditions for the line contact to investigate the fatigue life of OWC applications. This fatigue model is also used with CDM principles to study the subsurface-initiated damage.

As described earlier, a semi-infinite domain is considered for the analysis to reduce the computational time. Five different domains are randomly prepared and analysed to evaluate the initiation and propagation lives. As discussed in the previous section, the damage law for the elastic-plastic case depends on the incremental plastic strain. The stress histories obtained for the elastic-plastic model differ from the elastic model due to the involvement of micro-plasticity in the computational domain, and the cyclic response for case hardening steel is modelled using ELKP. The corresponding ELKP-based material model for the RVE under investigation has been defined in ABAQUS, as shown in **Figure 5.11**. For the contact analysis, a point in the subsurface level is considered for analysis where alternating maximum octahedral stress occurs. **Table 5.3** summarizes the FE simulation parameters, representing the material model and associated damage parameters.

Figure 5.11 ELKP material model in ABAQUS with UMAT

In order to implement the damage mechanics-based fatigue model, the damage evaluation laws along with damage coupled material constitutive equations require to be solved for each element simultaneously in the model to update the corresponding damage and elasticity modulus. Maximum contact pressure p_{max} varies from 3 GPa to 5 GPa, so a considerable amount of material undergoes plastic deformation. The friction coefficient value μ_s is 0.05 for elasto-hydro-dynamically lubricated surface conditions. Klecka et al. **[174]** suggested that the variation in modulus of elasticity is reported as small compared to residual and yield stress. So, the effect of case carburizing on the modulus of elasticity is neglected.

Table 5.3 FE simulation parameters

Parameters	**Value**
Modulus of Elasticity E, GPa	210
Poisson's ratio, $\boldsymbol{\vartheta}$	0.3
Contact pressure range, p_{max}, GPa	3-5
Yield strength, $\boldsymbol{S_y}$ GPa	2.054
Ultimate strength $\boldsymbol{S_{ut}}$, GPa	2.686
Fatigue resistance stress $\boldsymbol{\tau_{r0}}$, GPa	5.445
Damage rate exponent, m	9.87
Fatigue resistance stress for plastic law $\boldsymbol{S_0}$, MPa	38.3
Damage rate exponent for plastic law, q	2.17
Hardening Modulus M, GPa	9.53
Half contact width b, μm	200
Surface co-efficient of friction, μ_s	0.05
Critical damage parameter, D_c	0.95

The FE simulation results for OWC application obtained for five microstructure domains are investigated for fatigue life prediction for varying contact pressure values. Simulation parameters are kept identical for each domain. The damage progress from no-damage state to final failure remains unique for each domain due to distinct orientations and locations of Voronoi grains. **Figure 5.12** illustrates the damage evolution as a non-linear function of fatigue cycles for a randomly chosen element in the material domain. The damage increment value is varied from $\Delta D = 0.2$ to the critical value 0.95 (assumed). The damage evolution rate increases with a number of cycles, describing localized effective stress due to accumulated damage. The damage increment ΔD is chosen small enough to get the required accuracy of results but large enough to permit reasonable computational time.

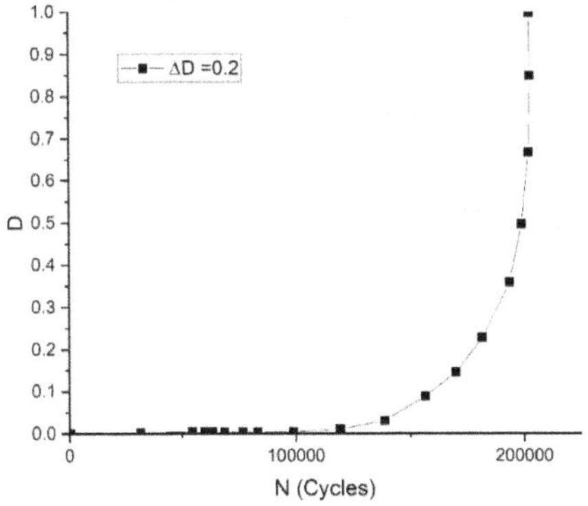

Figure 5.12 Damage evolution of randomly selected elements in the material domain

Figure 5.13 shows the equivalent contact stress induced at a subsurface level for 4 GPa contact pressure with available traction conditions for a single load cycle. The stress histories are utilized for the subsequent cyclic analysis for the numerical algorithm. For the line contact with a case hardening depth (CHD) up to 1.6 mm, the micro-plastic deformation only occurs at the case hardening layer. However, the crack initiation (about $390\mu m$) shows considerable deviation from the maximum octahedral stress depth (about $0.158\mu m$) for standard contact mechanics (elastic) where material strength, plasticity, or residual stress are not considered.

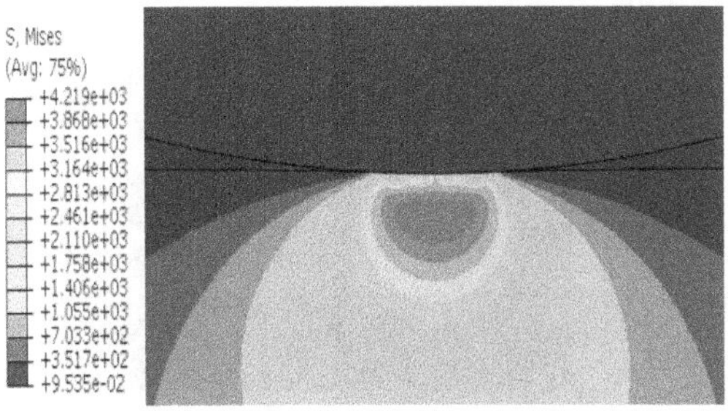

Figure 5.13 Maximum contact stress (MPa) induced for the contact pressure of 4 GPa

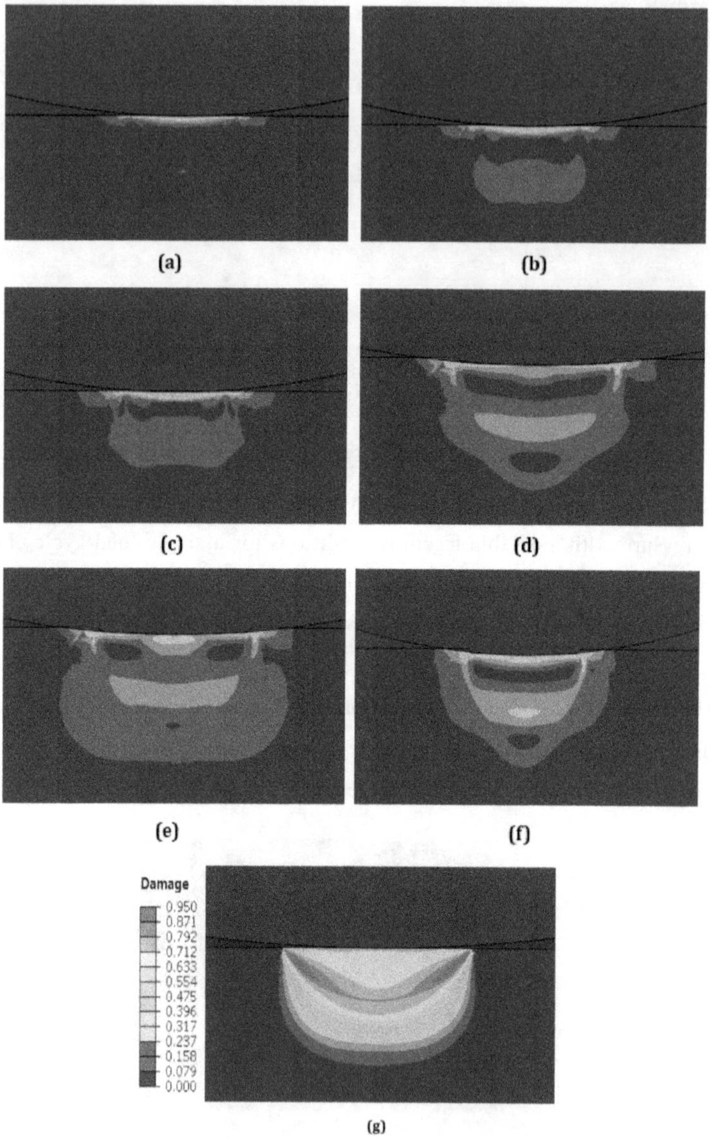

Figure 5.14 Damage evolution under simulation conditions of p_{max}= 4GPa and μ = 0.05 **(a)** Subsurface damage initiation at 164017 cycles. **(b)** Damage state at 174017 cycles. **(c)** Damage propagation to the surface at 180254 cycles. **(d)** Damage state at 182546 cycles. **(e)** Damage state at 185567 cycles. **(f)** Damage state at 189582 cycles. **(g)** Surface damage at critical damage state by 203912 cycles

Figure 5.14 illustrates the damage progression stages by a number of block cycles, starting from subsurface crack initiation to propagation. The multiple crack coalescence ultimately leads to the formation of the surface spall for critical damage value. The available surface traction also triggers the micro-crack initiation from the surface. Ultimately, the total life is evaluated for critical damage state by summation of crack initiation, and propagation lives for maximum contact pressure of 4 GPa at $\mu = 0.05$.

Figure 5.15 presents a graphical representation of fatigue lives obtained using an FE-safe coupled numerical algorithm. The worst life repeats are predicted for the critical damage state when the spall is developed at the surface. Fatigue wear due to smearing along with the subsurface-initiated cracks, which finally propagate to the surface, are responsible for generating such spalls. The failure of the OWC occurs when the contact surface wears out due to repeated loading-unloading, and the clutch is unable to transmit the required amount of torque.

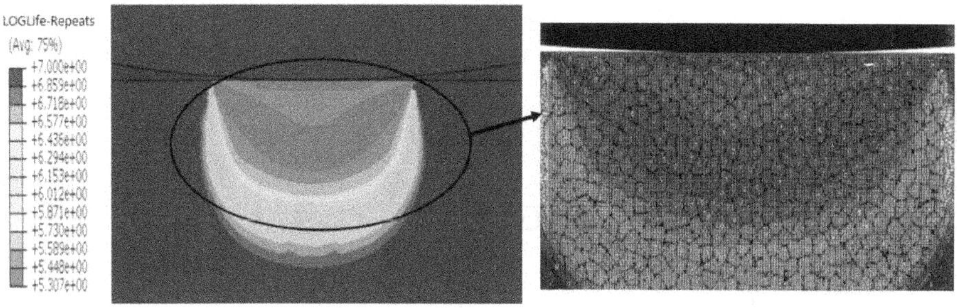

Figure 5.15 Fatigue life prediction corresponding to subsurface initiated spall under simulation conditions of p_{max}= 4GPa, CHD=1.5 mm, and $\mu = 0.05$

5.4.1 Effect of contact pressure variation on fatigue life

In order to investigate the effect of contact pressure variations on fatigue lives, all five microstructural domains previously employed are considered for the analysis of different contact pressure values. It is noted that the normal range of contact pressure for OWC applications is 3 GPa to 5 GPa. The derived L_{10} lives follow the inverse power law relationship with contact pressure as,

$$L_{10} = \frac{B}{{p_{max}}^{n}} \tag{5.20}$$

Where 'B' is the stress-life coefficient representing clutch geometry and material where 'n' is the stress-life exponent of the load-life relationship. However, such a nonlinear equation of L_{10} life depends on both B and n values to define their dependence on 'p_{max}'. The values of 'B' and 'n' for line contact models are derived by curve fitting using **Eq. (5.20)**. The stress life exponent for the L-P model is found to be 8.26, whereas for the current model, it is 9.136, which is closer to the material parameter 'm' considered for damage evolution law. It indicates the compatibility of torsion fatigue to contact fatigue for the current model and can be explicitly used to investigate the fatigue life of OWC for varying contact pressures with considerable accuracy. Similarly, the stress life coefficient is found to be 32675 and 64308, respectively. The five microstructure domains are analysed for fatigue life, and Weibull characteristics are obtained for the same. **Table 5.4** shows the life data and Weibull parameters obtained using the 2-parameter Weibull model. The results derived from the current model reveal that the elastic-plastic damage model predicts lives closer to the L-P model for low contact pressure. This is evident since plastic deformation is negligible at lower contact pressures than at higher ones. This means elastic stresses dominate life at low loads rather than plasticity-dominant loads at high contact pressures.

Table 5.4 Weibull statistics for numerical analysis

	Initiation Life			Total Life		
Scenario	**Weibull slope, e**	**Characteristic life, β (cycles)**	**L_{10} life (cycles)**	**Weibull slope, e**	**Characteristic life, β (cycles)**	**L_{10} life (cycles)**
p_{max} = 3 GPa, elastic-plastic	13.34	1979311	1005768	7.095	3867640	1683047
p_{max} = 4 GPa, elastic-plastic	14.38	190622	163008	6.129	294375	203912
p_{max} = 5 GPa, elastic-plastic	16.19	31012	26988	5.942	41609	28491

The Weibull probability plots produced from fatigue damage simulations measure the fatigue life and its scatter for different microstructure domains at varying contact pressures. **Figure 5.16** shows the data for the 2-parameter Weibull distribution characterized by L_{10} lives and their corresponding slopes. L_{10} measures life at 90% of the clutch population that has survived. The slope indicates a measure of scatter in life data, and the higher slopes give less scatter. It is because the material assumed in the simulation is free of inhomogeneity; only geometric changes in the microstructure model are dealt with for the analysis. A promising trend in fatigue life scatter is observed for the elastic-plastic model when pressure increases from 3 GPa to 5 GPa. The corresponding Weibull slope increases with pressure, which indicates decreasing scatter in initiation lives due to crack initiation dominance, and slopes decrease for final lives, which indicates high scatter since the failure becomes plastic strain driven.

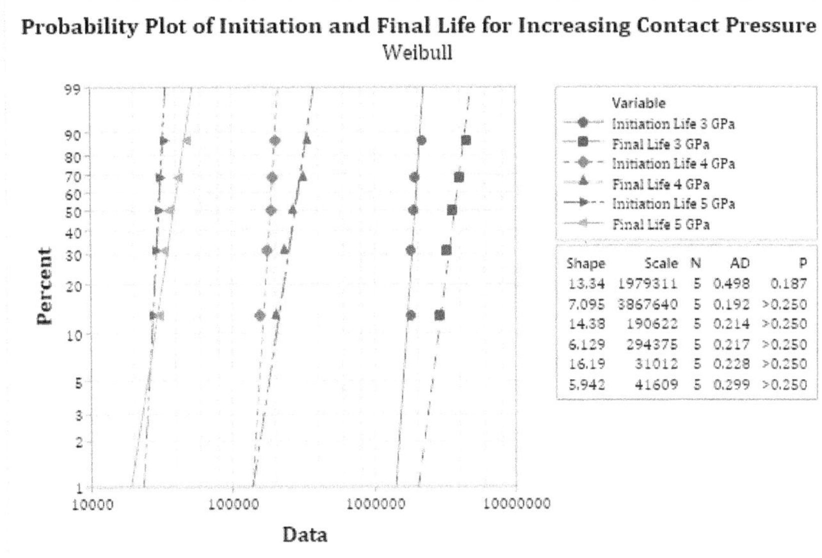

Figure 5.16 Weibull probability plots using the current model at different contact pressures (3GPa, 4GPa, and 5GPa) for 20MnCr5 material

The crack initiation and propagation life behaviour for increasing contact pressure values are shown in **Figure 5.17**. It is observed that for low-pressure values, i.e. 3 GPa, fatigue lives are mostly influenced by the propagation stage. The initiation stage contributes an average of 1/3 of total life. It means that initiation lives are less scattered and final lives are controlled mainly by the propagation phase. It is because, at low pressures, the failure is slowed down by anticipating the damage progression. L-P theory **[30]** also acknowledges that fatigue lives for bearing failures are crack propagation dominant.

Conversely, fatigue lives for higher contact pressure, i.e. 5 GPa constitutes a significantly shorter crack propagation phase. This is because crack propagation is significantly influenced by micro-plasticity in the heavily stressed material domain. Therefore, fatigue damage occurs due to accumulated plastic strains in the vicinity of the damaged intergranular joints that promote the majority of cracks to propagate further. Therefore, the crack propagation stage is very short, around 1/3 of the total life, and is the crack initiation dominant for heavily stressed contact pressures.

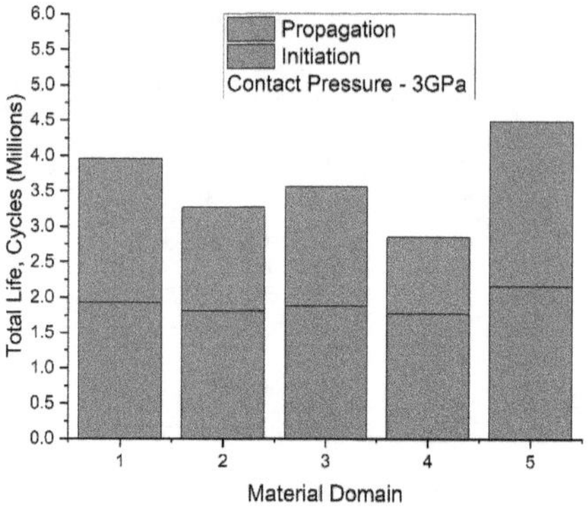

(a)

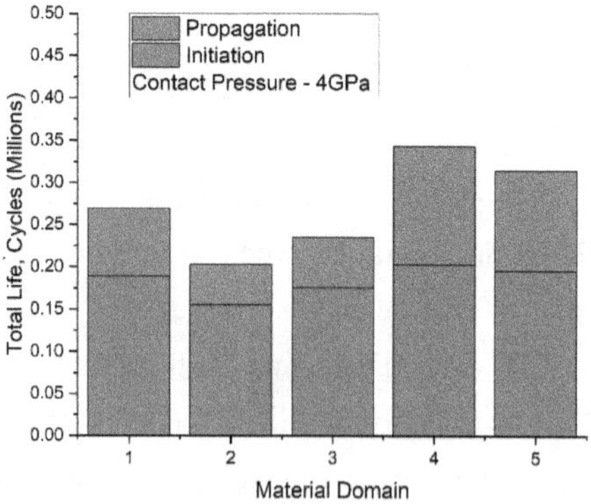

(b)

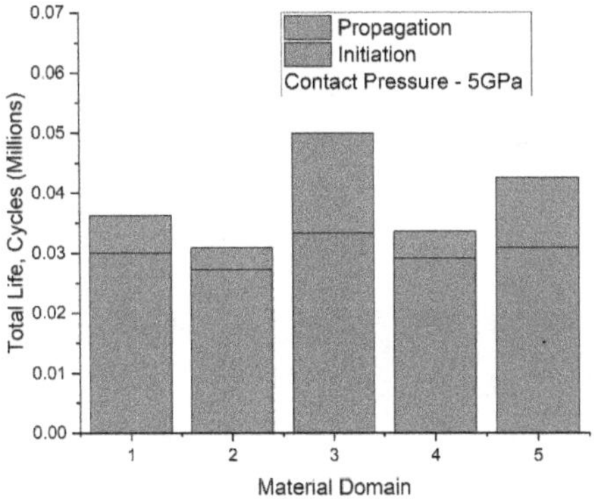

(c)

Figure 5.17 Crack initiation and propagation lives for contact pressures of **(a)** 3GPa, **(b)** 4GPa and **(c)** 5GPa using five randomly generated microstructures.

5.4.2 Effect of case carburization and residual stress on fatigue life

As discussed, the test specimens undertaken for experimental investigation were manufactured from case-carburized steel. The process of carburization results in a gradient in the material microstructure as a function of depth, which affects the mechanical properties of the material **[175-177]**. In order to study the effect of hardness variation, micro-indent tests are performed on the specimen, and the respective change in mechanical properties needs to be considered while material modelling. Similarly, the carburization process induces residual stresses in the case-carburized steel in the pre-fatigue state **[178-180]**. These residual stresses need to be evaluated and considered for numerical modelling to effectively predict fatigue life.

5.4.2.1 Measurement of effective case hardness using micro indent tester

The carbon content of a low carbon case hardening 20MnCr5 steel under consideration is increased by the gas carburization process to achieve fatigue resistance properties. The effective case hardness of the inner race section from the case to the core region is measured by the Vickers micro-hardness tester, as shown in **Figure 5.18 (a).** The typical case hardening depth for OWC varies from 0.8 mm – 1.6 mm for small to medium-sized specimens and extends up to 2 mm for large-sized specimens. **Figure 5.18 (b)** illustrates the indent marks for hardness variations measured with respect to the case depth. The first measurement is taken at 0.2 mm below the surface, followed by 25 readings each at ~0.2 mm apart by the micro-indenter. **Figure 5.18 (c)** shows that the readings closely follow the bilinear profile. The maximum hardness is measured at the surface, linearly decreases in the carburized region with case depth, and remains almost constant throughout the core. Thus, the case depth of the material is measured at 1.5 mm.

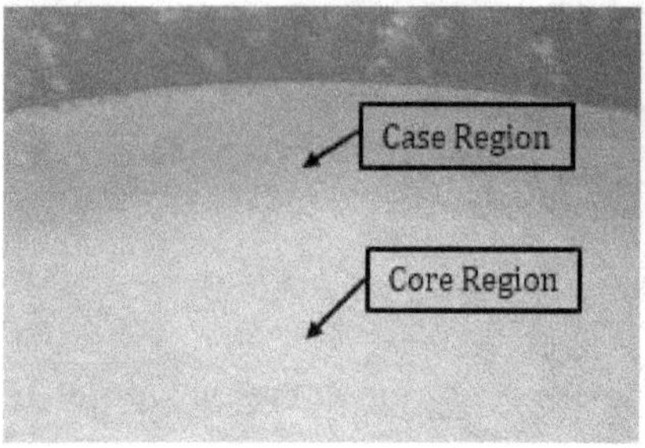

(a)

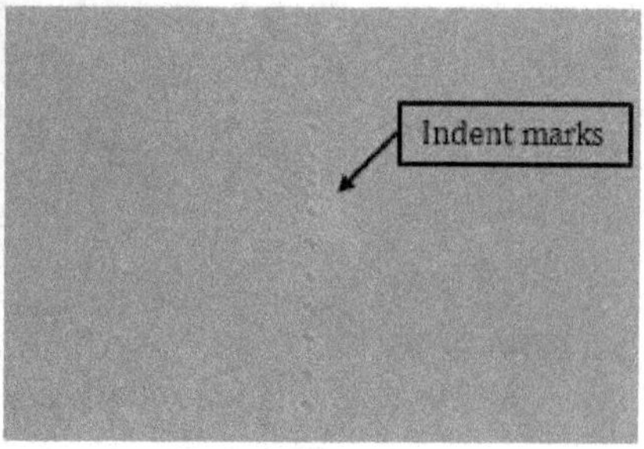

(b)

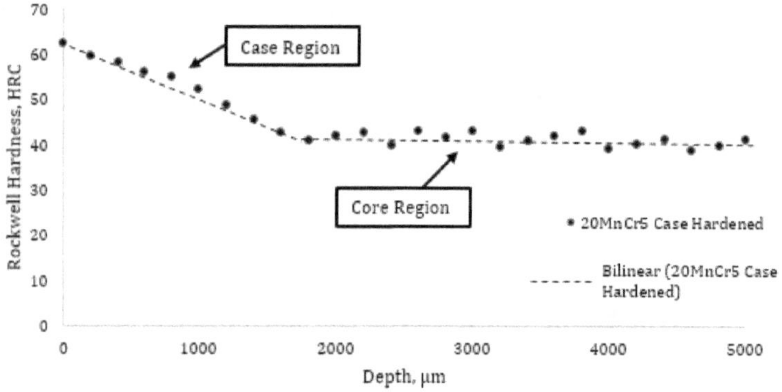

(c)

Figure 5.18 (a) Case and core regions of inner race section, **(b)** Micro-indent marks for effective hardness measurement from case to core regions, and **(c)** Vickers hardness measurement of case-carburized material specimen at varying depths from the surface to core region

For FEA analysis, three case depth values of 0.8 mm, 1.2 mm, and 1.6 mm with such a bilinear hardness profile are considered for evaluating the effect of case depths on fatigue life. It is accomplished by incorporating the hardness profile in the model itself by considering a linear relationship of yield strength (S_y) with hardness using **Eq. (5.3).** Therefore, any yield strength variation with case depth ultimately follows the corresponding hardness variation. **Figure 5.19** displays profiles of yield strength decay with case depths of 0.8 mm, 1.2 mm, and 1.6 mm, respectively, for the current investigation. The yield strength values in the simulation are assigned to each finite element under study with respect to case depths.

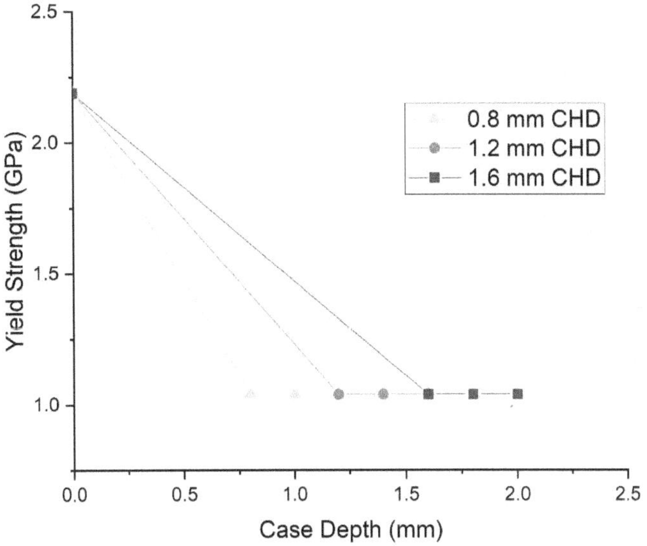

Figure 5.19 Yield strength variation with case depth for case carburized steel

5.4.2.2 Measurement of residual stress using X-ray diffraction (XRD)

An X-ray diffractography technique is used to study the atomic structure of the crystalline materials. Residual stress estimation using XRD technique involves the measurement of the strain in the crystal lattice of the material and its associated residual stress developed due to various machining and heat treatment processes. When X-rays impinge over the area of specimen where the stress is required to be measured, large number of grains and crystal particles contribute to the measurement. A crystal lattice consists of a periodic arrangement of atoms or ions. The planes of these atoms scatter X-rays, creating diffraction patterns that depend on the lattice structure. The intensity of the diffracted X-rays depends on the arrangement and type of atoms in the crystal. In addition, to yield strength variations with respect to hardness, a case carburizing process introduces compressive residual stress in the material. Compressive residual stress inside the material domain significantly improves the fatigue strength of case-carburized material **[181].** An X-ray Diffraction technique (XRD) in conjunction with the electro-polishing surface removal process is utilized for successive incremental material removal to measure residual stress variation with case depth. A clutch segment of the contact geometry has been considered for evaluating the compressive residual stress using an X-ray diffractometer, as shown in **Figure 5.20.**

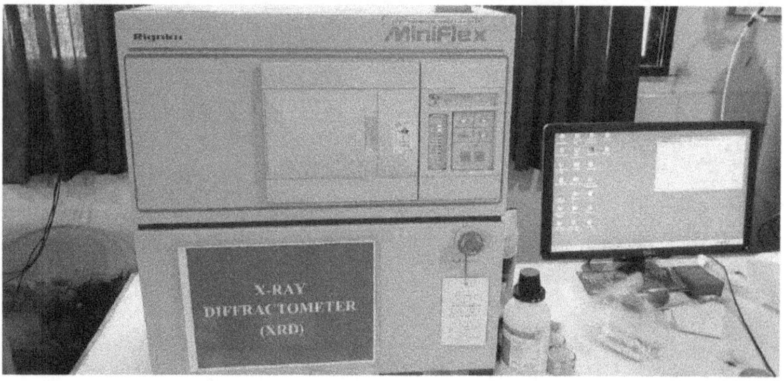

Figure 5.20 An X-ray diffractometer (XRD) for residual stress measurement.

An X-ray diffractogram plot obtained from X-ray diffractometer for the current analysis is depicted in Figure 5.21 (a). It consists of angle diffraction (2θ) measured in degrees on x-axis and the intensity of the diffracted X-rays measured in counts per second (CPS) on y-axis. The plot visually represents the interaction between X-rays and a material's crystal lattice, capturing the resulting diffraction pattern. This pattern provides information about the material's structure, phase composition, and other properties. Peaks in the diffractogram occur due to constructive interference from specific planes. These constructive interference occurs when the path difference between reflected rays satisfies the Bragg's law as expressed by **Eq. (5.21)**

$$n_d\lambda_a=2d_a\sin\theta \tag{5.21}$$

Where, n_d is the order of diffraction, λ_a is the X-ray wavelength, d_a is the spacing between atomic planes (inter-planar spacing), θ is the angle of incidence.

The residual stress is determined from the elastic constants (modulus of elasticity and poissions ratio) by assuming the linear elastic distortion of the crystal lattice plane. The most common method to evaluate the residual stress is $\sin^2 \Psi$ method, which correlates the lattice strain to residual stress. The lattice strain is plotted against $\sin^2 \Psi$, and by using the linear least square fitting, the slope of the line is determined for minimum R^2 value. Thus, the compressive residual stress is estimated using **Eq. (5.22) [182].**

$$\frac{1}{d_0}\frac{\partial d_{\Phi\Psi}}{\sin^2 \Psi} = \frac{1+\vartheta}{E}\sigma_{\Phi} \tag{5.22}$$

From the recorded peaks of XRD plots as depicted in **Figure 5.21 (a)** and by using **Eq. (5.22)**, the residual stress at the surface and peak value at critical depth evaluated for 1.6 mm case hardening depths (CHD) are -288 MPa and -430 MPa respectively. Similarly, the residual stress profiles for incremental CHD of 0.8 mm and 1.2 mm are evaluated and compared in **Figure 5.21 (b).** The experimental measurements of residual stress using the XRD technique indicate the maximum residual stress recorded at 20-25% of the value of d_s. Where d_s is the depth from the surface at which residual stress reduces to zero.

Residual compressive stress due to case carburization minimizes the magnitude of principal shear stress at the subsurface level. The residual stress effect is incorporated into the material model with corresponding case depth. Effective shear stress is evaluated by superimposing the residual stress on shear stress at the subsurface level. It is noted that the residual stress does not alter the principal normal stress in the radial direction. The residual stress variation obtained using XRD for corresponding case depth in the desired domain of the inner race is implemented in the model by assigning the corresponding values to each material point. It is noted that the hardness gradient in case hardening steel gradually reduces the fatigue strength, while residual stress present with increasing case depth improves the fatigue life. Therefore, a trade-off between these two phenomena must be considered to achieve maximum life.

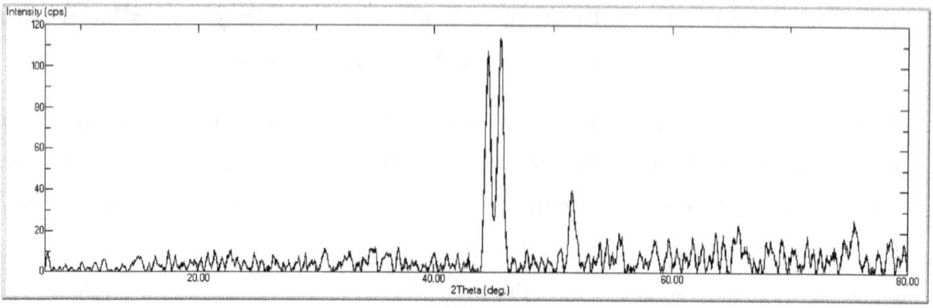

(a)

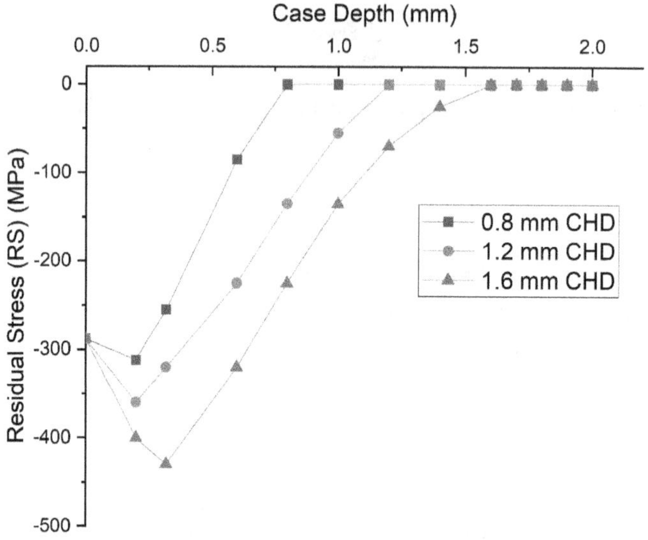

(b)

Figure 5.21 (a) Peak search graph using the XRD technique **(b)** Residual stress measured as a function case depth for 20MnCr5 case hardening steel

5.4.2.3 Fatigue life variation considering combined effects of yield and residual stress

In order to analyse the combined effect of yield strength and residual stress for case hardening steel, both yield strength and residual stress distribution are incorporated in the material domain with respect to case depths for performing the simulation. Weibull life plots are derived for case depths of 2b, 4b, 6b, and 8b for pressure values of 3 GPa,

4 GPa, and 5 GPa, respectively. It is observed that the combined effect of yield strength and residual stress results in the largest fatigue life for case depths of 4b at 3 GPa, 6b at 4 GPa, and 8b at 5 GPa. This is because the critical depth moves deeper inside the material at a higher amount of contact pressure. It is observed that fatigue life increases for increasing case depths for a fixed value of contact pressure up to a particular case limit, and then it decreases irrespective of further extension of case depth. It is because the values of damage parameters considered in the model vary with fatigue strength co-efficient and so with yield strength. So, the yield strength decreases with increasing case depth. The residual stress also increases up to the critical depth for the particular load value. Due to the simultaneous effect of both these parameters, an optimum case depth has to be defined at which maximum life can be achieved. Therefore, the optimum case depth for maximum life is the function of both yield strength and residual stress. The corresponding Weibull probability plots for different contact pressures are depicted in Figure 5.22, and respective Weibull parameters are displayed in **Table 5.5**.

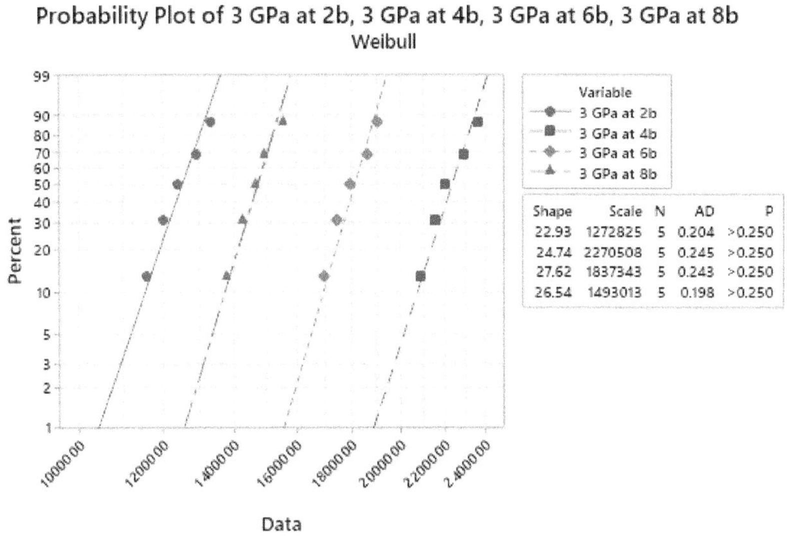

(a)

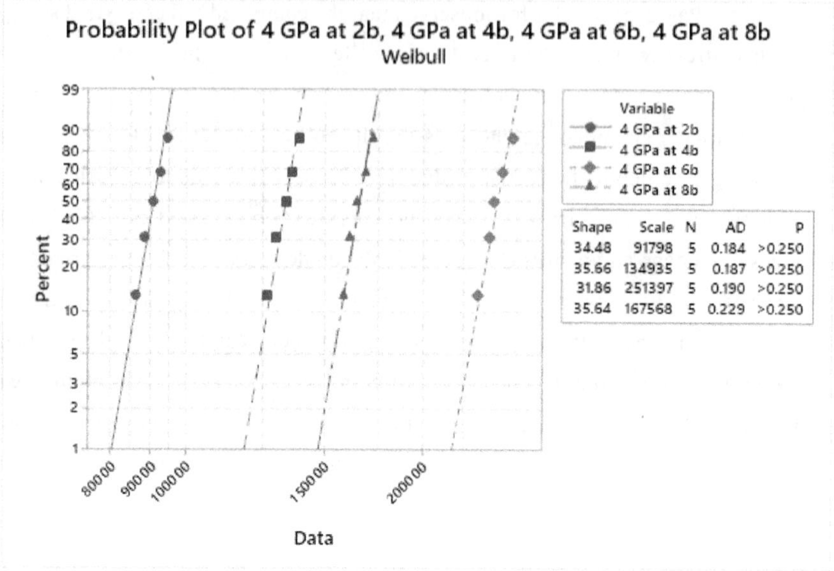

(b)

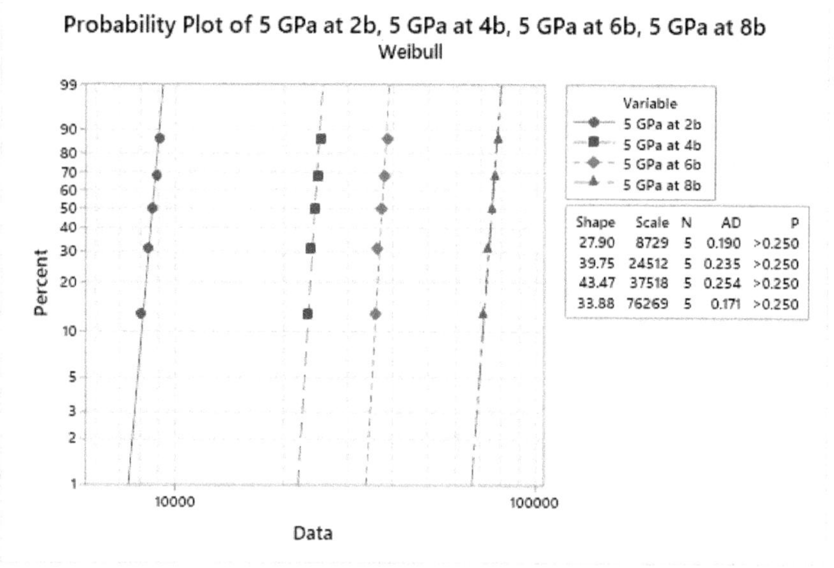

(c)

Figure 5.22 Weibull probability plots for various case depths at contact pressures of (a) p_{max} = 3GPa (b) p_{max} = 4GPa (c) p_{max} = 5GPa

Table 5.5 Weibull statistics for varying case depths

Scenario	CHD	Weibull slope, e	Characteristic life, β (cycles)	L_{10} life (cycles)
p_{max} = 3 GPa, elastic-plastic	2b	22.93	1272825	1153843
	4b	24.74	2270508	2073096
	6b	27.62	1837343	1693580
	8b	26.54	1493013	1371637
p_{max} = 4 GPa, elastic-plastic	2b	34.48	91798	85998
	4b	35.66	134935	126683
	6b	31.86	251397	234253
	8b	35.64	167568	157315
p_{max} = 5 GPa, elastic-plastic	2b	27.90	8729	8053
	4b	39.75	24512	23163
	6b	43.47	37518	35625
	8b	33.88	76269	71368

5.5 Numerical Fatigue Life Estimation using CDM and Life Comparison

An experimental test rig is specifically developed to simulate the required function of OWC with unidirectional repeated loading for estimating fatigue lives, as described in Chapter 3. Accelerated life tests are conducted to expedite the failure, thereby shortening the life of components. The fatigue life data are recorded at higher torque levels and extrapolated to the operating levels using an appropriate load-life relationship, as discussed in Chapter 4. The failure of the component is considered when micro spalls due to sub-surface initiated cracks, which, in addition to surface traction, cause the amount of material liberated from the contact surfaces. Such shallow scar or spall on the contact surface ultimately results in a steeper contact angle between rollers and inner race, and the available clamping friction is insufficient to transmit the required torque, and eventually, failure is observed. The number of cycles to failure is recorded using a PLC-based load measuring system.

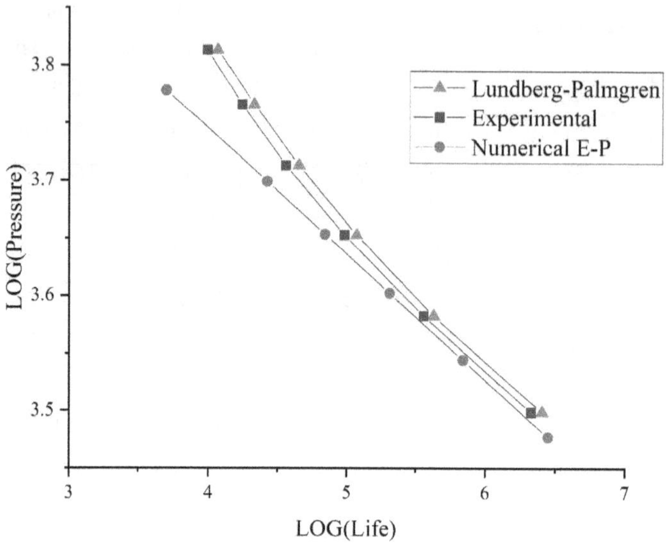

Figure 5.23 Comparison of fatigue life curves for case hardening 20MnCr5 steel for OWC.

Figure 5.23 illustrates the variation in fatigue life with contact pressure using the current model. An effort is made to correlate these results with the experimental and L-P model. The experimental results obtained using accelerated tests for five samples are utilized to develop an experimental load-life relationship using a 2-parameter Weibull estimate to account for the scatter in fatigue lives. Similarly, the numerical life results for five distinct microstructure domains are used to generate the numerical load-life relationship for the current model. The fatigue life results obtained using the elastic-plastic model in log scales show deviation from other models but follow a good trend with them at lower ranges of contact pressure due to negligible plasticity. The variation in fatigue life results is observed due to the assumptions made in the L-P model. The L-P model is based on an empirical relationship that includes geometrical and material constants. The L-P model fails to consider lubrication film, available surface traction, and surface-initiated failures like scratches, dents and flaws. However, in the current model, the surface traction effects are considered, which push the maximum orthogonal shear stress region nearer to the surface, thereby accelerating the failure and delivering lower lives compared to the L-P model.

Furthermore, the results collected using this model reveal that the elastic-plastic model predicts fatigue lives closer to the lives obtained using the L-P model and the experimental load-life relationship at lower contact pressures. As the contact pressure increases, a significant difference in fatigue life is observed due to the percentage of life consumed being more in crack propagation than crack initiation, and the effect of plasticity dominates at higher loads, which accelerates the damage. Therefore, the numerical results show a considerable reduction in life results at higher loads than other models. Similarly, the experimental results obtained in accelerated conditions also have their assumptions of extrapolations. Due to all these reasons, the fatigue lives predicted using the current model differ from the theoretical and experimental values.

5.6 Summary

In this investigation, a damage mechanics coupled finite element model is developed to predict the fatigue life of OWC. The numerical simulation evaluates crack initiation and propagation lives for the defined contact surfaces. Thus, the total numerical fatigue life of the clutch is the summation of lives estimated during the crack initiation and propagation stages. Furthermore, the material microstructural randomness and material plasticity have been taken into account by the Voronoi tessellation-based elastic-plastic material model. In order to consider the effect of case carburization on fatigue life, micro indent tests are performed to examine the change in mechanical properties with respect to case depth. XRD tests also evaluate the compressive residual stress due to case carburization. The fatigue lives obtained under these combined effects indicate that the life increases for increasing case depths for a fixed value of contact pressure up to a particular case depth limit, and then it decreases irrespective of further extension of case depth. That means the optimum case depth for maximum life is the function of both yield strength and residual stress. The fatigue life results obtained using the elastic-plastic model show deviation from other models but follow a good trend with experimental results at lower ranges of contact pressure due to negligible plasticity.

Chapter 6#

Experimental Wear Life Estimation of OWC using Pin-On-Disc Tribometer and their validation using Finite Element Model.

This chapter presents an experimental set-up for wear measurement of OWC under freewheel mode using a pin-on-disc tribometer. The experimental test plan and test methodologies are discussed to estimate the wear life of OWC. An FE-based numerical model is proposed to validate the experimental results, and the wear life of OWC is estimated with respect to sliding speed.

GRAPHICAL ABSTRACT

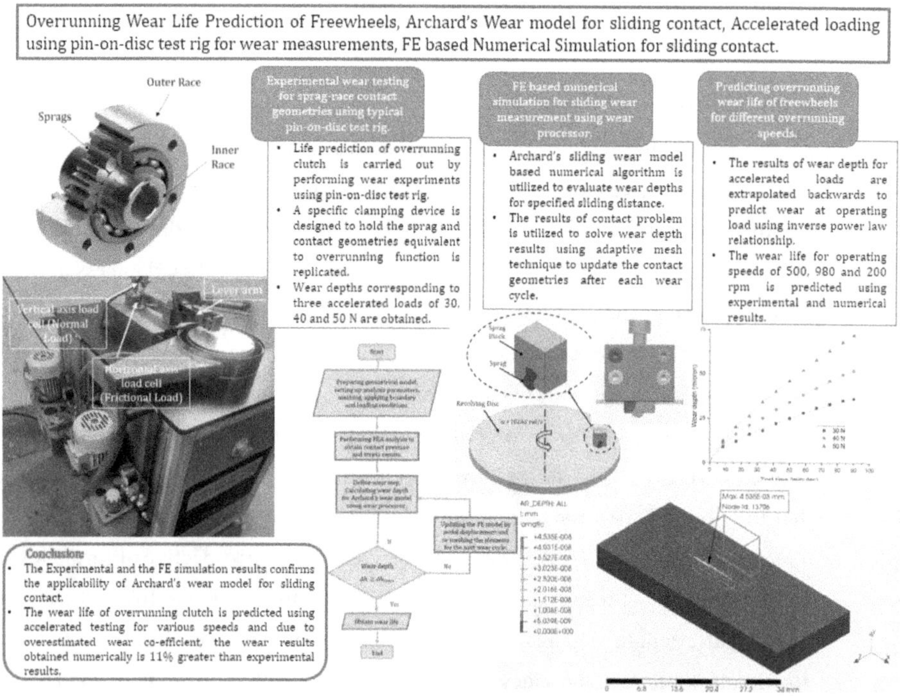

#*Karan A Dutt*, *Shashikant J Joshi, Dhaval B Shah and Dipak Prajapati (2024). "Experimental investigations and finite element simulation for predicting wear life of overrunning clutches." Engineering Research Express. Volume 6, Number 3, 035522.*

Sliding wear involves removing the peak asperities of the surface not only due to abrasion but also due to other mechanisms involved, such as adhesion, subsurface crack initiation and progression, diffusion of the material in contact, etc. **[183-186].** Archard was the pioneer to propose a linear wear equation for characterizing sliding wear generally expressed by **Eq. (6.1) [187]** as,

$$Q = k\frac{F_N.\, s}{H} \tag{6.1}$$

Where 'Q' is the wear volume (mm^3), 'F_N' is the normal load (N), 's' is the sliding distance (mm), 'H' is the hardness of the softer material (N/mm^2), 'k' is the wear coefficient.

The wear coefficient is the probability of dislodging the wear particle or debris from the material surface and depends on the surface roughness, asperity interface, and the amount of loading **[188].**

6.1 Experimental Wear Measurement Set-up using Pin-on-disc Tribometer.

A conventional pin/ball on disc tester (Model No: TR-20LE- PHM400-CHM400, Make: Ducom Instruments Private Limited) **[189],** as shown in **Figure 6.1,** is a typical floor mounting test rig provides the facility to study the wear and friction characteristics of the test specimen under dry, lubricating and heating conditions. It works on the principle of pressing the stationary pin/ball on the revolving disc to measure the sliding wear of materials in contact. It also assesses the material characteristics of the various tribo-systems in laboratory conditions. The normal and frictional loads, along with rotational speed and wear track diameter, are the variables that need to be specified for specific tests. The standard mechanical and testing specifications of the test rig are listed in **Table 6.1.**

The test rig consists of the control unit and the test unit. The control unit is operated by a PLC (programmable logic control) based controller interfaced with a monitoring device. The control unit is a combination of the electronics and instrumentation controller, which is interfaced with the computer through a signal output cable. It is equipped with a hydro-pneumatic system to operate the mechanical devices installed within the test rig.

Figure 6.1 A conventional pin-on-disc tribometer for wear measurement. (Courtesy: Nirma University, Ahmedabad.)

The test unit is mounted on the base plate, which consists of a rotating disc and specimen holder facilitated by a lubrication cup, as shown in **Figure 6.2.** The sensors associated with the load, speed, temperature and wear measurements are part of this testing unit. A lubrication recirculation unit of 3 litres capacity and a gear pump are utilized to provide controlled boundary lubrication with a flow rate of around 250 ml/min, nearly exhibiting similar operating conditions.

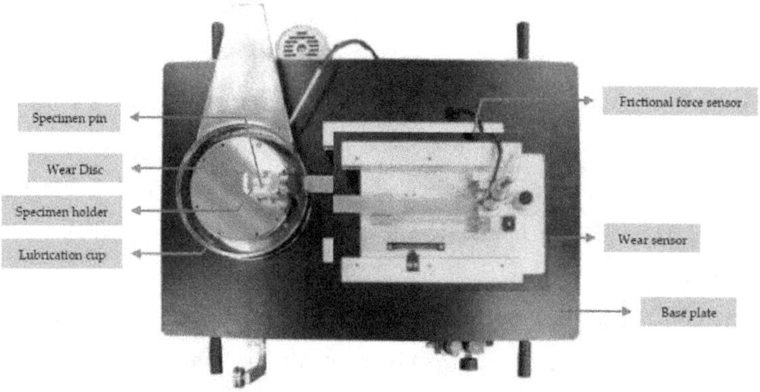

Figure 6.2 A control unit and the test unit of standard pin-on-disc tribometer **[189].**

Further, the operation of the test rig is controlled automatically by user-oriented software running with the Microsoft operating system. The parameters for the required operation are assigned to the software, and the output from the sensor is displayed and recorded on the software.

Table 6.1 Standard mechanical and test specification of pin-on-disc test rig **[189]**

Part	Specification
Specimen	Pin size: - Ø3, 6, 10 & 12 mm; (For ambient & heating condition) Ball size: Ø10mm; (For ambient & heating condition)
Specimen holder	Split hard jaws (ambient condition) and collet (for heating condition for clamping Ø3, 6, 10 & 12 mm pins; and Ø10 mm ball.
Wear disc size	Diameter 165 mm x 8 mm thick; Material: EN 31 hardened to 58-62 HRC
Wear track diameter	Min: 30 mm , Max :135 mm
Disc Speed	Range: 200 to 2000 rpm Sensor specification: Proximity sensor Least count: 1 rpm Accuracy: (1 ± 1% of measured speed) in rpm
Frictional force	Range : 0 – 200 N Sensor specification: Load cell Least count: 0.01 N Accuracy : (0.01 ± 1% measured frictional force)in N
Wear	Range: Up to ± 2000μm Sensor specification: LVDT Least count: 1 micron Accuracy : (1± 1% of measured wear) in microns
Normal load	Range: 20 – 200 N (Pneumatic loading) Sensor specification: Load cell Least count: 1 N Accuracy : (0.1 ± 1% measured frictional force)in N
Temperature	Ambient-400˚C (Pin Heating and chamber heating)
Lubrication unit	The lubrication re-circulation unit has a gear pump, pressure setting valve, valve to discharge oil 100 -750 ml/min, and oil tank 3-litre capacity.

6.1.1 Materials and contact geometries

A typical pin-on-disc tribometer usually offers a sliding contact, making a surface-to-surface contact during wear measurements. Conversely, the overrunning clutch involves sliding sprag elements on the race surfaces by introducing a line contact along their length during the overrunning operation. In order to replicate a similar line contact interface on the existing test rig, it is required to develop a specific clamping device as shown in **Figure 6.3 (a),** which can hold the sprag in such a position on the disc that a line contact between two sliding surfaces is achieved. In other words, the original test rig is modified by replacing the pin holder with a specifically designed clamping device to hold the sprag on the disc to attain an overrunning function.

Figure 6.3 (b) illustrates the schematic of contact geometries for sprag-on-disc tests. The materials of sprag and disc tested for wear experiments are listed **[10,168]** in **Table 6.1,** along with their mechanical properties. In the present study, linear elastic constitutive models for both the sprag and the disc are considered for the analysis. The materials for both disc and sprag are thorough-hardened steel with hardness values of 54 HRC (Hardness Rockwell C scale) and 62 HRC, respectively.

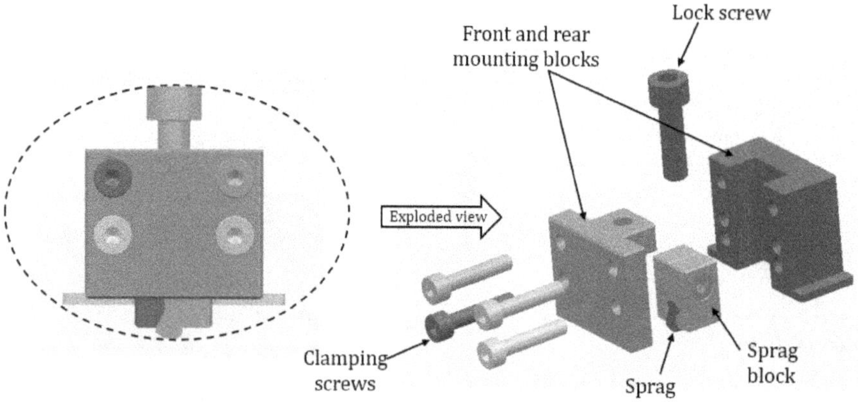

(a)

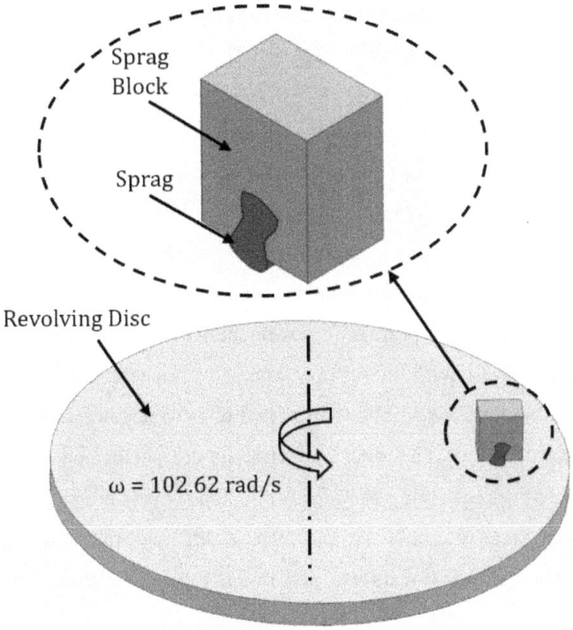

(b)

Figure 6.3. **(a)** A schematic of a sprag holding device (clamping device) with an exploded view **(b)** A schematic of sprag-on-disc contact geometries.

Table 6.2. Material properties of sprag and disc **[10, 167]**

Material	Chemical composition (wts %)	Hardness (HRC)	Roughness Ra (mean) (μm)	Density (kg/m^3)	Elastic Modulus (GPa)	Poisson's Ratio
Sprag DIN 20MnCr5	0.2 C - 1.032 Cr - 0.3 Si - 1.21 Mn - 0.029 P - 0.017 S	62	0.175	7850	210	0.3
Disc AISI 52100	1 C - 1.031 Cr - 0.35 Si - 0.45 Mn - 0.026 P - 0.037 S	54	0.204	7810	210	0.3

6.2 Test Plan and Methodology

In order to investigate the wear characteristics of overrunning clutches, a pin-on-disc test rig is utilized to perform wear tests for sliding contact between sprag and disc with a thin film of lubricant (Mobil) having a grade of DTE 24 (dynamo turbine engine). Test samples are prepared according to ASTM (American Society for Testing and Materials) G99-05 standard **[190]** and cleaned before each test run to remove any debris. The experiments are performed by assigning normal load values as input along with the track radius and speed at which the disc is revolved. During overrunning operation, a tiny amount of frictional force is induced at contact surfaces in the presence of lubrication. Progressive wear is also not that significant when measuring with such low frictional force. It may take thousands of hours to achieve the allowable wear limit of 200 μm, i.e. the maximum permissible wear between sprags and race, below which the clutch functions well without failure. Therefore, accelerated tests are performed with more severe loading conditions than actual ones to accelerate the amount of wear with a short testing period. The wear experiments are performed under elevated load levels rather than at the operating load. Elevated load levels are selected in such a way as to accelerate the wear rate only without introducing any other effect or change in the wear mechanism. Qualitative tests are performed to confirm these accelerated levels without introducing any other effect that would never occur during normal operating conditions.

A good test plan for conducting wear experiments consists of the selection of load levels, operating conditions and the number of test specimens on which the accelerated testing should be carried out. For the present investigation, three elevated load levels (30 N, 40 N, and 50 N) are considered with a constant track radius of 42.5 mm for collecting the wear data at higher load levels and backwards extrapolating these data for estimating the wear amount at operating load level (4 N). Such incremental loads are selected to analyse the wear progression at contact assembly. Five sprag specimens are considered at each of these load levels, which makes it a total 15 number of wear measurements required to be carried out to map the accelerated model prominently. Based on that, a test matrix is prepared, as displayed in **Table 6.3,** to perform the wear experiments effectively and record the wear data to estimate the wear life of the clutch. In addition to the selection of load levels, the accelerated tests have been designed to simulate the several hours of wear into 1.5 hours of test runs with realistic limits of the material under test.

Table 6.3 A test matrix for conducting wear experiments

Use Condition 4 N	Elevated Load Conditions		
	30 N	40 N	50 N
Sample Unit			
1	38	55	79
2	38	50	78
3	36	53	76
4	35	54	73
5	34	49	71

For bearing steel under investigation, the loading values have thus been selected by considering all these factors. The diameter of the disc is 162 mm with 6 mm of thickness with track radius of 42.5 mm. Hence, the sliding distance covered is calculated as 23500 m for 4.36 m/s sliding velocity at 980 rpm. The wear depth up to ±2000 μm is measured using an LVDT (linear variable differential transformer) based vertical axis load cell by means of the lever arm in the normal direction. The frictional force in the tangential direction is measured using a horizontal axis load cell in the range of 0-200 N on the same lever. The hydro-pneumatic operated lever arm continuously maintains the contact of the sprag with the rotating disc during tests by constantly applying the normal load. The pin-on-disc rig, along with the specimen clamping device, is shown in **Figure 6.4.**

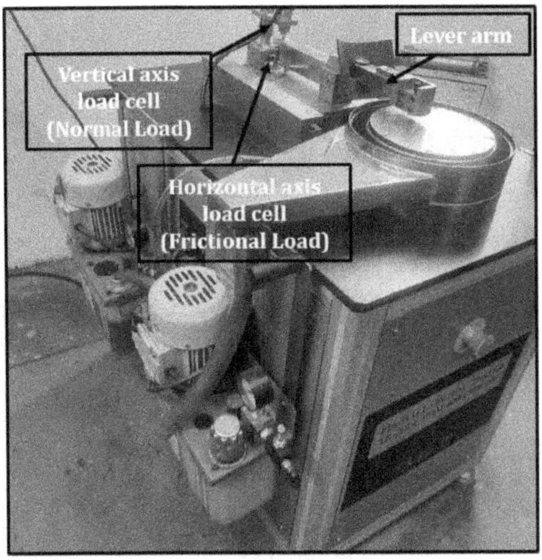

(a)

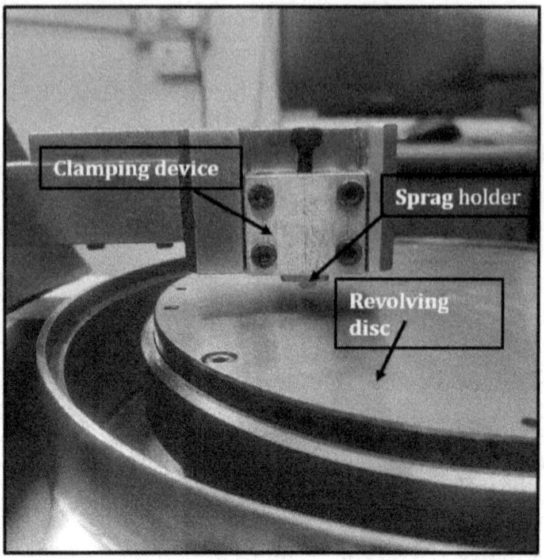

(b)

Figure 6.4 (a) A typical pin-on-disc wear test apparatus for sliding wear measurements **(b)** A clamping device to hold the sprag on the disc.

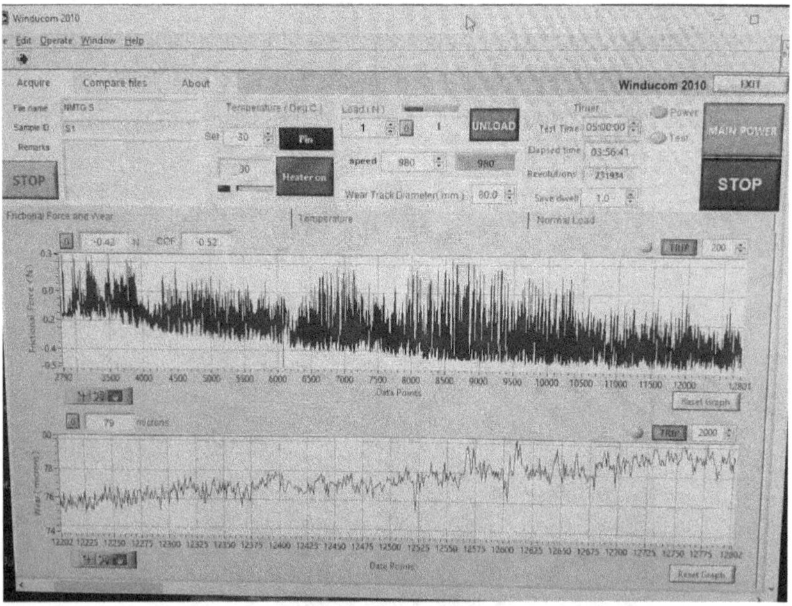

Figure 6.5 A graphical user interface to operate the tribometer with defined testing parameters

Test output is in the form of GUI (Graphical User Interface), as displayed in **Figure 6.5,** which shows the friction and wear patterns recorded during the defined period for the defined test parameters as input. The input parameters, such as the values of normal load, test duration, the rotational speed of the disc, and the wear track diameter, are fed into the system for the defined test module. The system returns the values of measured wear depth and friction coefficient as output for the set input parameters. The test trial can also be terminated in case of any emergency by pressing the "stop" button. At the end of the testing, the user can avail the raw data generated for each time step in the desired file format and use them for further investigations.

6.3 Experimental Results and Discussion

Figure 6.6 (a) illustrates the experimental wear results of sprag-disc materials with respect to test time with lubrication under three accelerated loads of 30, 40 and 50 N. It indicates that the wear rate is initially high due to the contact area being small in earlier stages of wear, which are called running-in testing periods **[191-194].** The material is removed gradually as wear progresses and the actual area of contact increases, resulting in a decrease in the contact pressure and wear coefficient, which, thereafter, achieves a steady state condition. The allowable limit for wear depth is 200 μm beyond which, if the clutch is operated, wedging between the sprag and race does not occur, leading to overturning the sprag in the opposite direction while transmitting the torque, and the clutch fails. For the current study, it is not necessary to run the test for such a long time to achieve this limit; instead, all the experiments are carried out with a fixed test duration of 1.5 hours. The wear life of the clutch at operating conditions is thereafter predicted by extrapolating these results using the appropriate load-life relationship observed during analysis **[195].** The experimental wear data obtained for three load levels are plotted for the definite test periods. These data have been represented by a curve fitting technique, i.e., power law fit, available in the data analysis software. Consequently, the generated curve has been extrapolated to estimate the wear life for the permissible wear depth value, i.e. 200 microns, as illustrated in **Figure 6.6 (b).**

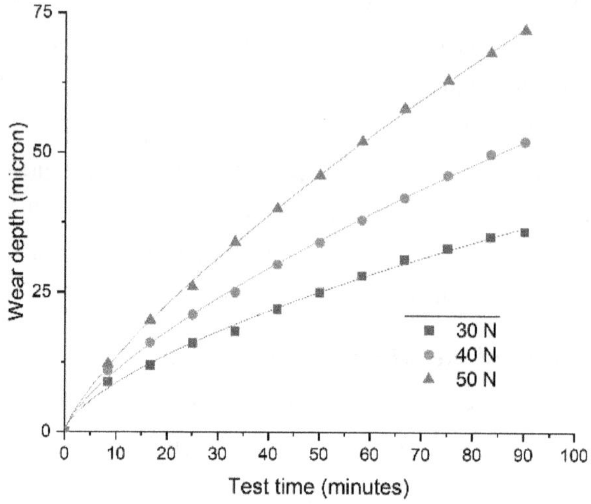

(a)

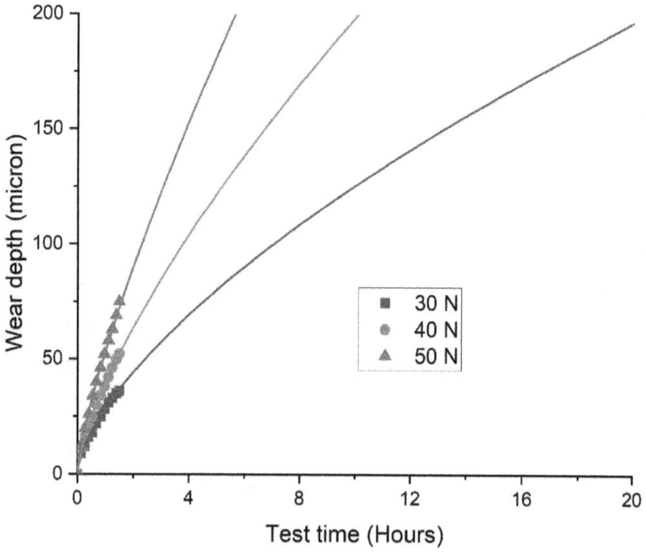

(b)

Figure 6.6 (a) Experimental wear results from the sprag-on-disc tribometer at three accelerated normal loads **(b)** Extrapolated model relationship for wear depth limit of 200 μm under three accelerated normal loads.

The frictional characteristics like frictional force and frictional torque are continuously monitored during tests via an LVDT-operated lever arm in the tangential direction. The coefficient of friction values is determined by knowing the applied normal and frictional forces. **Figure 6.7** depicts the variation of friction coefficient with sliding distance for material under tests for three loads. It is observed that the friction coefficient initially increases and achieves a steady state due to the formation of wear debris and wear of surface asperities. The mean (average) value of the coefficient of friction extracted from the experiments for the three load levels is considered 0.085**[10]**.

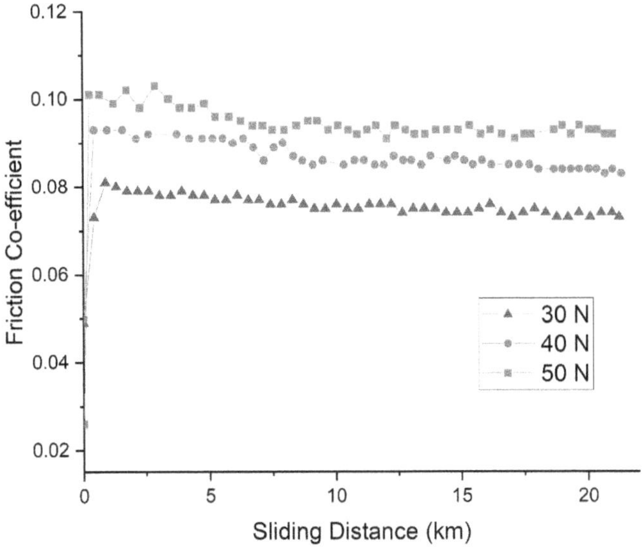

Figure 6.7 Friction coefficient profiles experimentally obtained for accelerated loads of 30 N, 40 N and 50 N with respect to sliding distance.

The wear coefficient under three load tests is also measured for materials under contact, which falls under the category of mild wear. The wear experiments are performed for five samples, each with three accelerated load values. The repeatability of the results is confirmed by plotting the results of the experiments using normal distribution. The mean values for wear depths are calculated, and the scatter of the results is plotted in terms of probability plots for normal distribution, as illustrated in **Figure 6.8**. **Table 6.4** shows the corresponding values of wear depth, wear rate and friction coefficient obtained for all three loads.

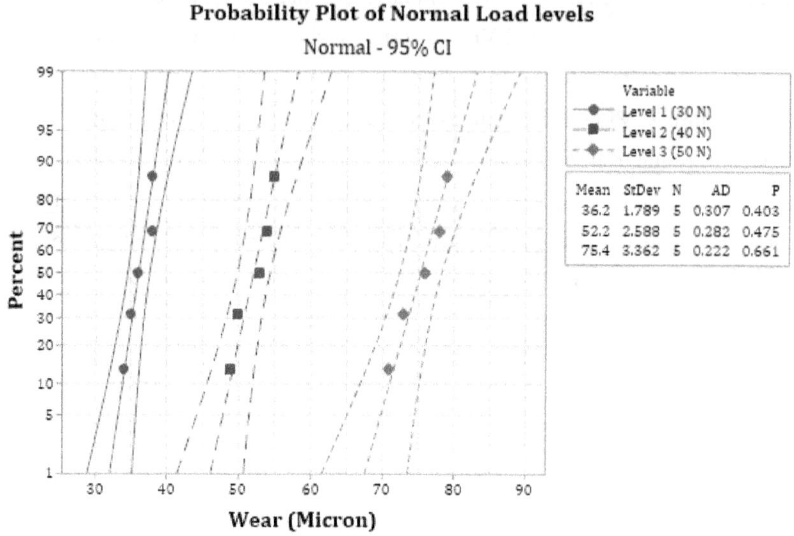

Figure 6.8 Probability plots of experimental wear depths for normal loads (30 N, 40 N and 50 N) with five samples each.

Table 6.4 Experimental outcomes, test run = 1.5 hours

Normal Load (N)	Wear depth (μm)	Wear rate (mm^3/m)	Friction coefficient
30	34	2.85E-10	
30	35	2.93E-10	
30	36	3.02E-10	0.075 ± 0.01
30	38	3.19E-10	
30	38	3.19E-10	
40	49	4.75E-10	
40	50	4.85E-10	
40	53	5.14E-10	0.085 ± 0.01
40	54	5.23E-10	
40	55	5.33E-10	
50	71	7.69E-10	
50	73	7.91E-10	
50	76	8.24E-10	0.095 ± 0.01
50	78	8.45E-10	
50	79	8.56E-10	

The wear regimes have been observed through qualitative tests before actual test runs. The range of applied normal load for the current testing and operating conditions has been identified first, i.e. 20 N to 50 N under 1.5 hours of the testing under controlled lubrication. The wear regimes have been identified based on the wear rate and characterized by the comparative amount of wear regimes, as depicted in **Figure 6.9**. The present analysis describes wear regimes in terms of the comparative amount of wear obtained. Mild wear is observed below 30 N load values with the lowest wear rate, whereas severe wear is above 40 N load values with the highest wear rate, separated by a transition regime from mild to severe when the load ranges from 30 N to 40 N with wear rate observed in between the lowest and highest values.

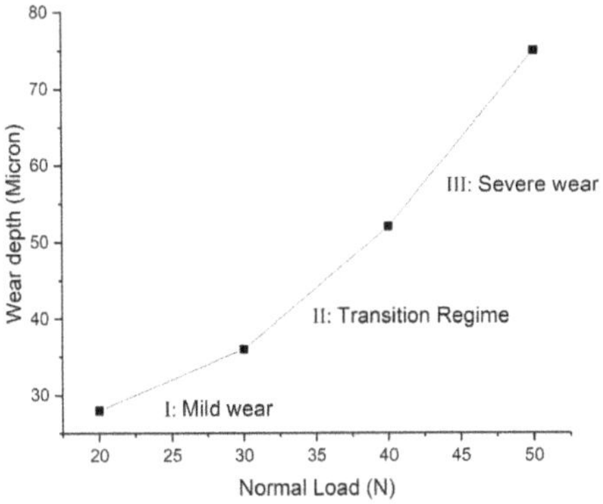

Figure 6.9. Wear regimes recognized corresponding to normal loads.

6.3.1 Drag torque measurement of OWC

The frictional torque of freewheels in overrunning operation is specified as drag torque. It is the torque required to rotate one of the races for freewheeling operation. Since the sprag comes into contact with the inner race due to the spring, the drag torque is determined by the spring force. Spring force is relatively small, and sprags rotate at the same speed as the outer race. When the outer race rotates, centrifugal force acts on sprags; thus, the drag torque also varies according to the rotational speed of the race. The drag torque measuring load cell, as shown in **Figure 6.10,** is utilized to measure the available drag torque of the clutch. It is noted that the value of drag torque varies with available track radius, surface roughness and lubricating conditions.

Figure 6.10 A drag torque measuring load cell. (Courtesy: NMTG Private Limited)

Technically, a drag torque per 1 kgf of load on the clutch is the frictional torque applied at the contact of sprag elements for a force of 1 kgf (10 N) applied over a distance arm of 1 mm/minute for the OWC sectional geometry as shown in **Figure 6.11**. It can be expressed by **Eq. (6.2)**

$$T_D = \mu \times F_t \times \text{length} \quad (6.2)$$

$$= 0.085 \times 1 \text{ kgf tangential load (10 N)} \times \text{Couple (2 x inner radius)}$$

Where 'T_D' is the drag torque at the contact of sprags and inner race, 'F_t' is the tangential force at the sprag raceway contact, and 'μ' is the coefficient of friction range for OWC.

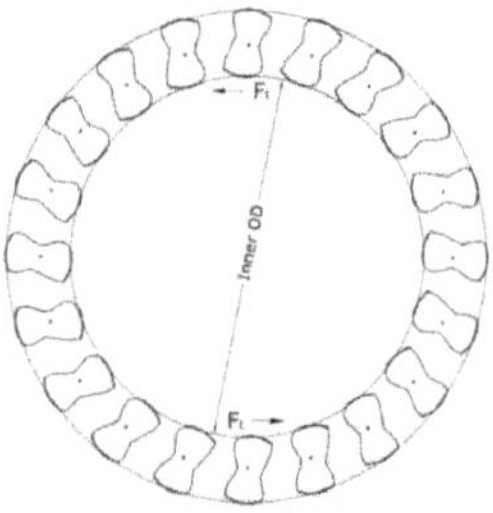

Figure 6.11 Drag torque calculation from clutch sectional geometry

Moreover, as discussed, the contact force between the sprags and the inner race changes under the influence of centrifugal force caused by the rotation of the sprag cage assembly. In this context, the OWC can be of engaging and disengaging types based on the drag torque variation with speed. The contacting force between sprags and inner race reduces (disengaging)/ increases (engaging) according to the centrifugal force. The rotational speed at which the drag torque becomes zero in disengage type clutch and the relevant speed is called disengage speed, above which the sprags do not make contact with the inner race. Since the engaged type clutch increases the drag torque with the rotational speed of the outer race, sprags positively engage with inner and outer races when it change to engagement during the outer race rotation.

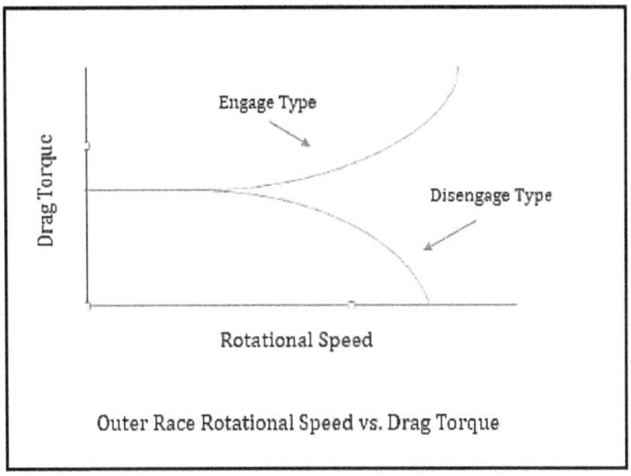

Figure 6.12 Drag torque variation with rotational speed.

On the other hand, the disengaged type clutch decreases the drag torque with the increase in rotational speed of the outer race, which reduces the wear on the sprags and the inner race contact surface. However, the disengagement type clutch is not suitable for engagement in High-speed applications. A graphical representation of the relationship of drag torque with rotational speed is illustrated in **Figure 6.12.**

6.3.2 Load-life relationship

In a current study for the track radius of 42.5 mm, the average value of drag torque per sprag is measured as 16.25 N-mm. The corresponding frictional and normal force per sprag for 0.085 friction coefficients are calculated as 0.36 N and 4 N, respectively. So, the wear results for this load indicate the wear life of the overrunning clutch at nominal conditions. In this way, the wear depth adjacent to the 4 N normal load is predicted without performing the experiments for a long time using the inverse power law relationship from the results of three accelerated loads, as illustrated in **Figure 6.13**. Thus, the overrunning wear life of freewheel running at 980 rpm is evaluated by extrapolating this relationship backwards from the accelerated results. In addition, the Weibull analysis for estimating the life characteristics at the operating load level is performed for the available wear data for three elevated load levels of 30 N, 40 N and 50 N, respectively. The Weibull probability plot of experimental wear data for five samples, each under three load levels, has been displayed in **Figure 6.14,** and the life characteristics for operating load level, i.e. 4 N, has been evaluated using appropriate load-life relationship.

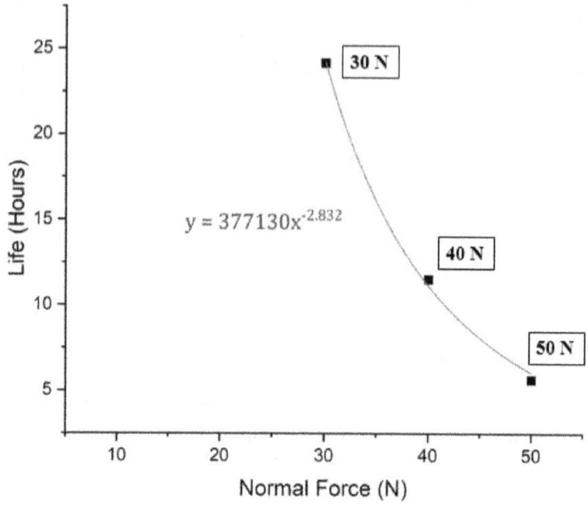

Figure 6.13 Load-life relationship derived from the experimental wear results for three accelerated loads of 30 N, 40 N and 50 N.

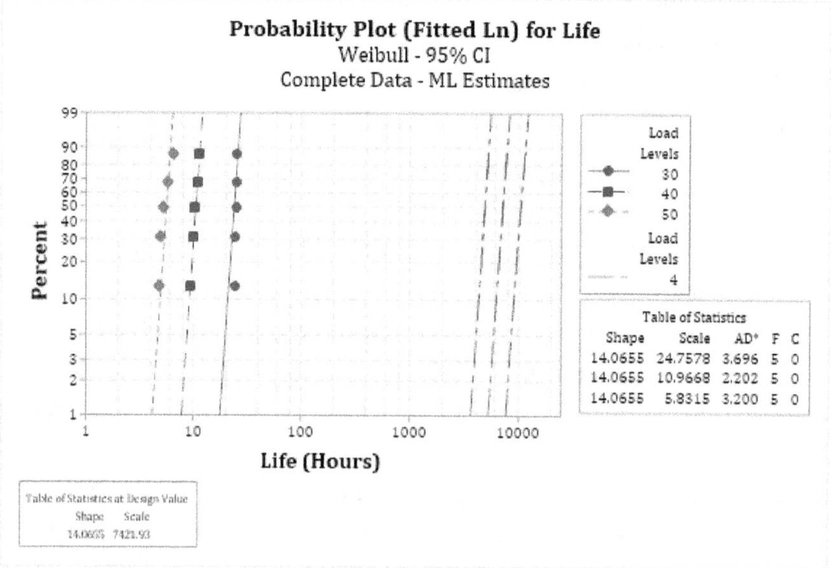

Figure 6.14 Weibull probability plot for estimating life characteristics at operating load level (4 N) from experimental wear results for three accelerated loads of 30 N, 40 N and 50 N.

6.4 Finite Element-based Numerical Model for Wear Life Estimation

A finite element model is prepared to represent the contact geometries of the experimental work using any of the CAD software. The CAD model is utilized to run the boundary and loading conditions as per the experimental test parameters. The numerical wear values under these conditions are predicted and validated by the experimental results to predict the numerical wear life.

A 3-dimensional CAD (computer-aided design) assembly of sprag-on-disc geometries has been prepared as shown in **Figure 6.15 (a)** for performing numerical simulation and is found consistent with the features of the experimental test setup. Instead of selecting the whole body for the simulation, only a small portion of the model is taken into consideration for performing a wear cycle of 25 mm sliding distance. The geometrical parameters used for preparing the FE model are listed in **Table 6.5**.

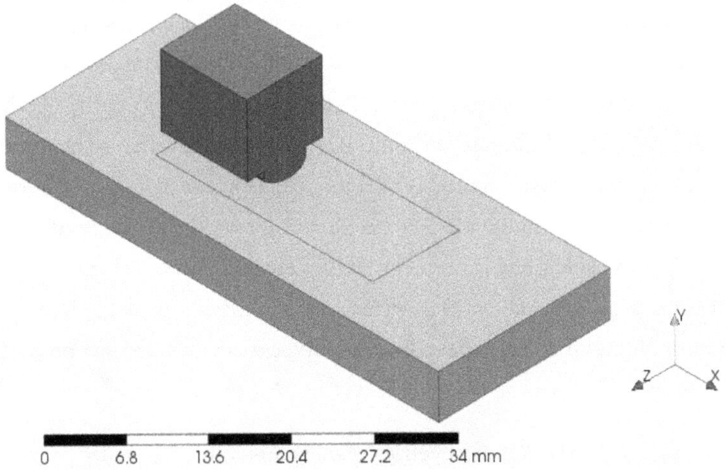

(a)

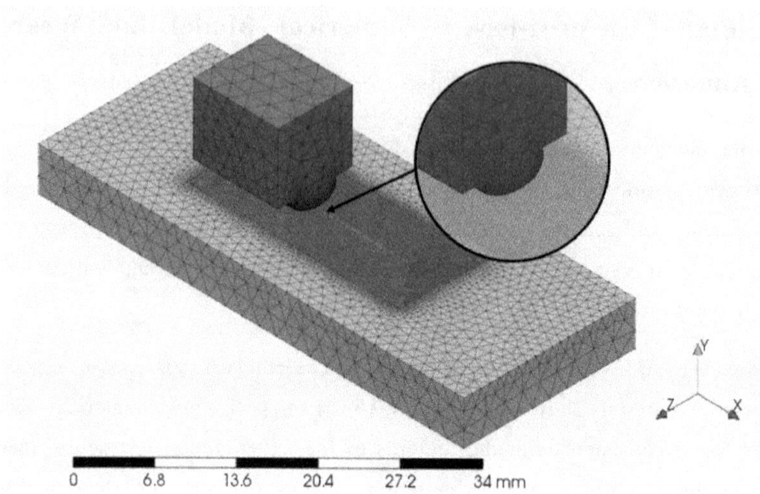

(b)

Figure 6.15 (a) A solid model for sliding contact assembly **(b)** An FE meshed model.

A 3-dimensional solid model is finely meshed at the contact region and coarsely meshed at the other locations away from the contact to obtain accurate results of the contact problem, as shown in **Figure 6.15 (b).** The element type selected for both the sprag block and the plate is linear tetrahedral C3D4 with a total number of elements obtained as 13980 and 90834, respectively, for 0.2 mm element size. A mesh convergence study was performed to decide the element size and improve the accuracy of the numerical results. The mechanical and material properties are assigned as described in section 6.1. The thermal aspects at the contact surface are assumed to be negligible in the model since no meaningful changes have been observed in the mechanical properties due to these effects.

Table 6.5 Geometrical parameters of the FE model

Parameter	**Value (mm)**
Sprag height, d_s	8
Sprag length, l_s	10
Plate dimensions, $l_p \times w_p \times t_p$	50 × 20 × 5
Sprag block dimensions, $l_b \times w_b$	8.5 × 10

A numerical approach for evaluating the wear life of OWC is illustrated in **Figure 6.16** in terms of a flow diagram. A FE analysis has been carried out using the wear processor (PrePoMax) **[196]**. The solid geometry has been imported into the processor, and the relevant boundary and loading conditions are applied in the processor itself, which reflects a GUI interface window similar to ABAQUS.

Figure 6.16 A flow diagram for evaluating wear life using a numerical approach.

Wear simulation is executed in two steps: the first involves the contact analysis, and the second involves computing wear from the solution of the contact problem based on Archard's equation. The loading and boundary conditions are applied to the model as illustrated in **Figure 6.17** based on the experimental conditions described,

1. Three accelerated normal loads of 30, 40, and 50 N are applied on the top surface of the sprag block for both steps.
2. A linear displacement of the sprag block in the y-direction is allowed for the contact step, and a linear displacement of 25 mm is applied in the x-direction for the wear step. All other displacements and rotations of respective steps are constrained.
3. All displacements and rotations of the plate are restricted (the plate is kept fixed).
4. The friction and wear coefficient values are considered as input parameters to the FE model.

The material model for executing the contact analysis is prepared, and the corresponding wear step for the current analysis is defined as shown in **Figure 6.18** for the available data using the wear processor.

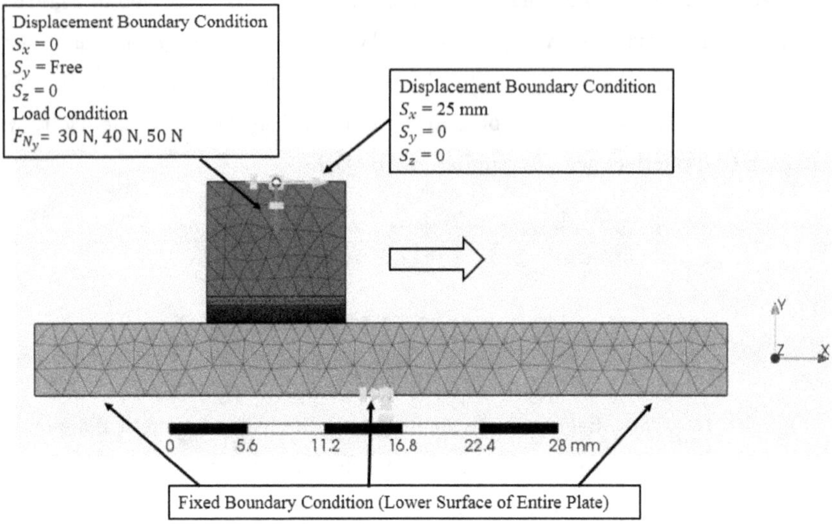

Figure 6.17 Boundary conditions imposed on the FE model.

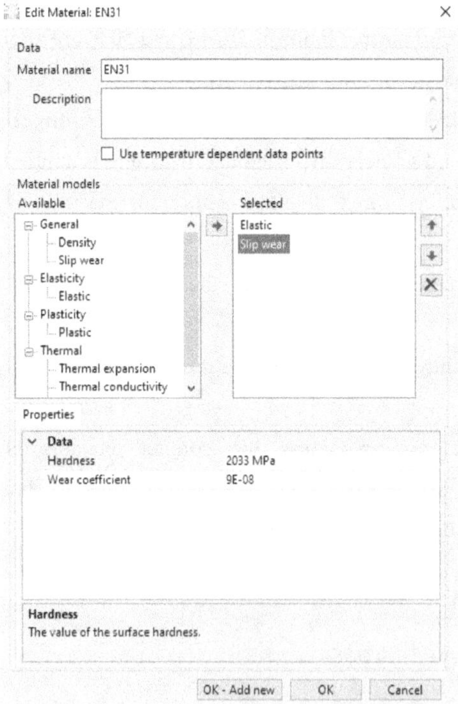

Figure 6.18 The wear step defined for the FE model using the wear processor.

6.4.1 Simulation strategy for wear estimation

The FE wear processor implements Archard's wear model to evaluate sliding wear at every node of the contact surface with every wear cycle by updating the contact surface profile. The wear depth at each contact node of the FE model is evaluated using Euler's explicit integration method. The average contact pressure 'p' is evaluated as the ratio of the applied normal force 'F_N' to the available contact surface area. From **Eq. (6.3) [122]**, it is evident that for infinitesimally small contact area 'ΔA', the wear depth increments 'dh' perpendicular to the contact interface with a sliding distance 'ds' can be expressed by **Eq. (6.4) [122]** as,

$$\frac{dQ}{\Delta A} = dh = k\frac{F_N}{\Delta A}\frac{ds}{H} \tag{6.3}$$

$$\frac{dh}{ds} = \frac{k}{H}p \tag{6.4}$$

Thus, the incremental wear depth over sliding distance is calculated by nodal displacements at the local scale using Euler's integration method based on Archard's wear model by **Eq. (6.5)-(6.6),**

$$h_{j+1} = h_j + \Delta h_j \tag{6.5}$$

$$h_{j+1} = h_j + Kp\Delta s_j \tag{6.6}$$

Where K= k/H is the specific wear rate (mm^2/N), p is the average pressure (MPa), h_{j+1} is total wear for $(j+1)^{th}$ step and h_j is worn up to the jth step.

The wear processor utilizes an adaptive mesh technique for updating finite element mesh due to nodal displacements after each wear cycle to modify the contact surface due to surface wear. A simplified flow diagram for the wear simulation strategy is presented in **Figure 6.19.** After solving equilibrium equations and achieving convergence for the contact problem under imposed boundary conditions for the first step, the wear step is invoked by the processor to calculate the wear depth for all contact nodes. The wear processor then provides all the arguments required to apply the adaptive mesh technique for modifying the contact surface for the next wear cycle with updated contact pressures. The model does not directly consider the effect of the debris available at the contact surface. The friction coefficient and wear coefficient values obtained from the experiments are utilized as input parameters in the processor. The wear depth limit Δh_{limit} is defined for the clutch beyond which, if operated, the clutch would not be able to perform its function correctly and fails due to loss of clamping friction. The wear processor stops running the simulation once the allowable wear limit is achieved, and the overrunning wear life is predicted based on the wear amount.

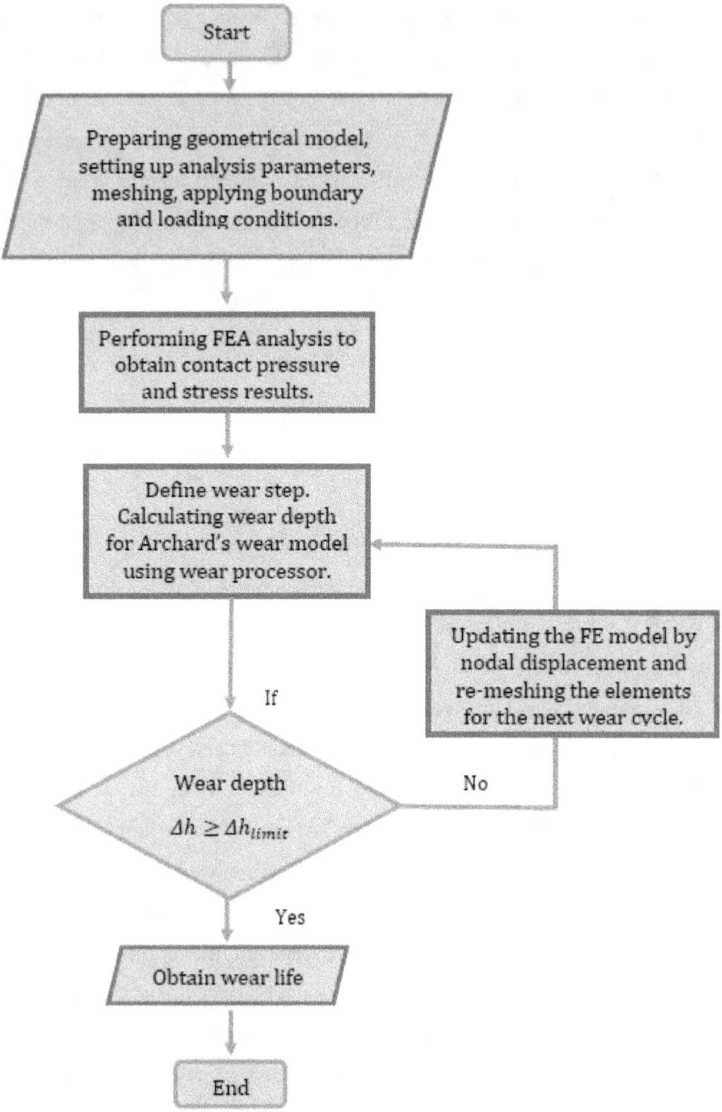

Figure 6.19 Wear simulation flowchart of sliding contact for overrunning clutches.

6.5 Numerical Results and Discussion

A finite element wear processing tool is utilized for wear prediction based on Archard's wear model. Using Euler's integration scheme, the local wear at the contact regions is computed and integrated over the sliding distance as described in section 6.4.2. The wear processor progressively performs a series of static simulations by updating the surface geometries at the contact surfaces for particular loading conditions. The deformed geometries due to wear after every wear step are re-meshed, ensuring uniform meshing for further computations. Since experimental wear tests are conducted for accelerated load conditions with a short test run, the coefficient of friction curves initially shows high values due to running-in periods **[197]**. However, numerical analysis is performed with the friction coefficient value of the steady state period only **[198]**. The Hertzian pressure distribution at the contact region is defined as a surface traction vector by **Eq. (6.7) [11]**,

$$F_y(x) = p_{max}\sqrt{1 - \left(\frac{x - x_c}{b}\right)^2} \quad (6.7)$$

Where $F_y(x)$ is the normal component of contact traction, p_{max} is maximum pressure at the centre, (x_c, y_c) is the coordinate of the centre at maximum pressure, and 'b' is the half-contact width.

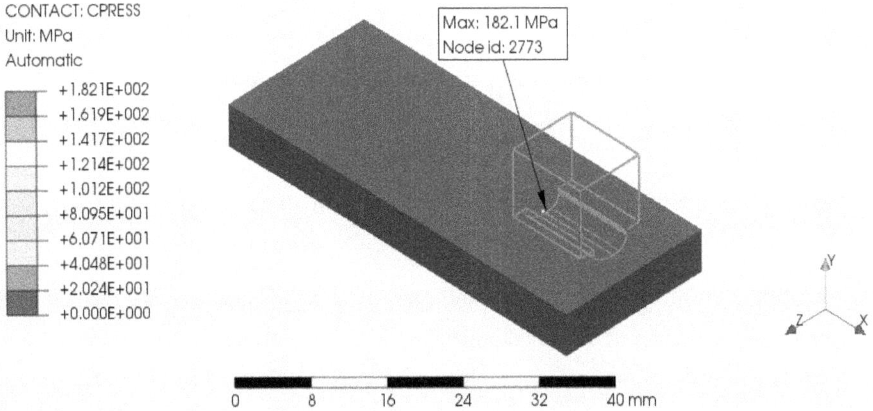

(a)

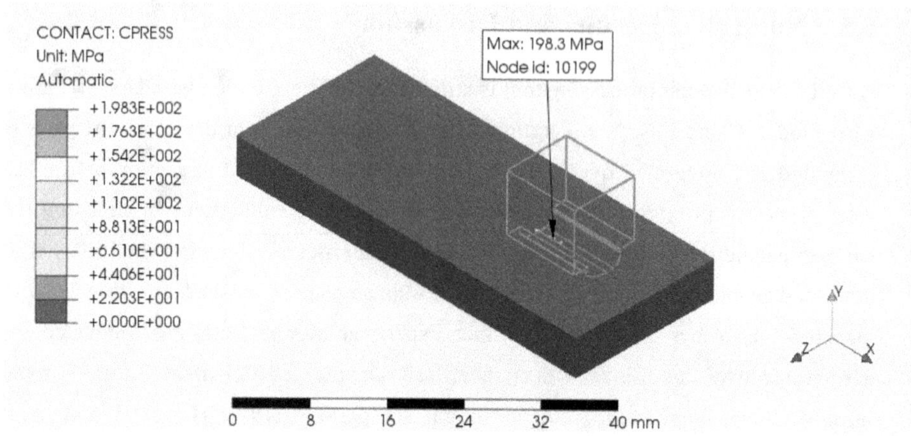

(b)

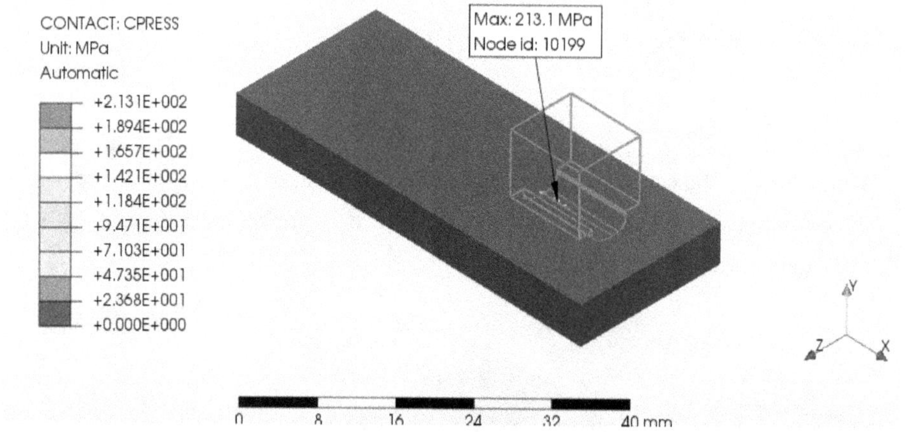

(c)

Figure 6.20 Numerical contact pressure for wear cycle of 25 mm sliding distance at **(a)** 30 N, **(b)** 40 N and **(c)** 50 N load values.

As a first step of the numerical solution strategy, an FE analysis for evaluating the contact surface under defined loading and operating conditions is illustrated in **Figure 6.20.** The solution of the contact problem using the FE model indicates good co-relation with Hertz contact theory for 30 N load, as shown in **Figure 6.21 (a)** in terms of contact profiles. Therefore, it is ensured that the current FE model can be explicitly used to estimate the wear. The number of iterations for numerical simulation has been carried out with different element sizes. The element size of 0.2 mm has been finally achieved to improve the numerical results that closely converged with Hertz's closed-loop solution for the contact model. The corresponding mesh convergence study has also been performed and illustrated in **Figure 6.21 (b)** with respect to element size for your reference.

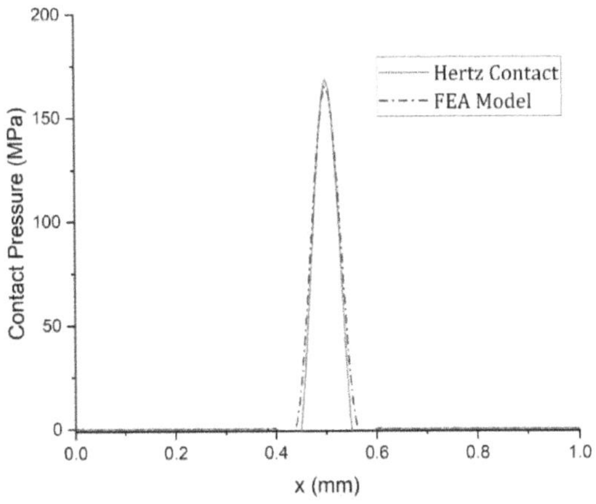

(a)

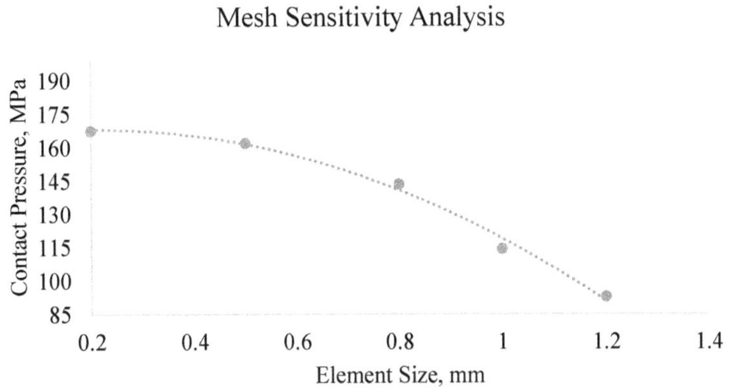

(b)

Figure 6.21 (a) Contact pressure distributions for FE model compared with Hertz contact model for 30 N load. **(b)** Mesh sensitivity analysis

Figure 6.22 illustrates the wear depth contours simulated for a 25 mm sliding distance under 30 N, 40 N and 50 N normal loads, respectively. The maximum local wear for corresponding contact pressure values 167.8 MPa is computed as 4.53 x 10^{-08} mm for the contact between sprag and disc using experimentally obtained wear and friction coefficient as inputs. Similarly, the wear depths for 40 N and 50 N normal force are obtained as 6.32 x 10^{-08} and 8.92 x 10^{-08} for contact pressure values of 179 MPa and 214.5 MPa, respectively.

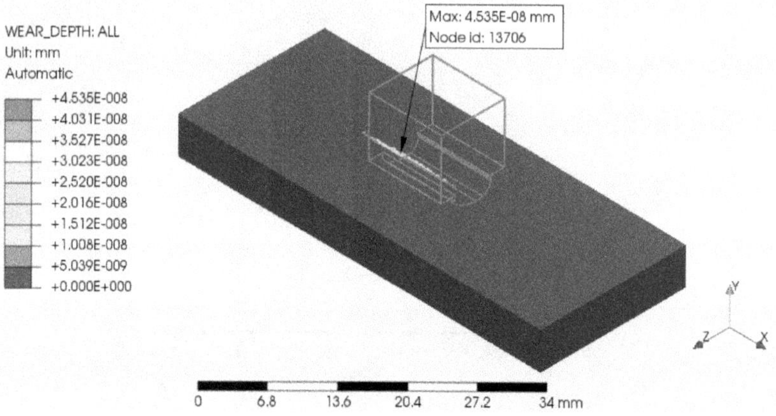

(a)

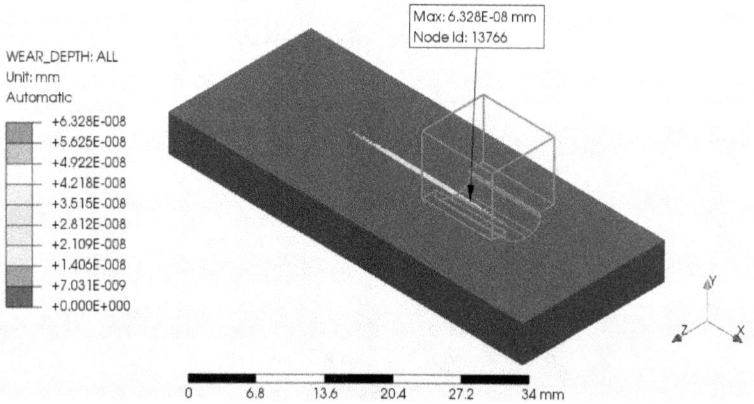

(b)

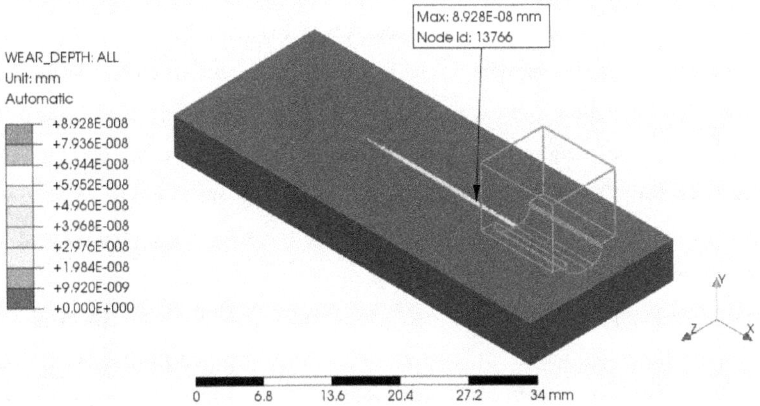

(c)

Figure 6.22 Numerical wear results for wear cycle of 25 mm sliding distance at **(a)** 30 N load, **(b)** 40 N load, **(c)** 50 N load

Figure 6.23 presents the wear simulation results of sprag-disc materials derived by running the simulation for 3000 mm of sliding distance under 30 N, 40 N, and 50 N of normal loads, respectively, as per Archard's wear model.

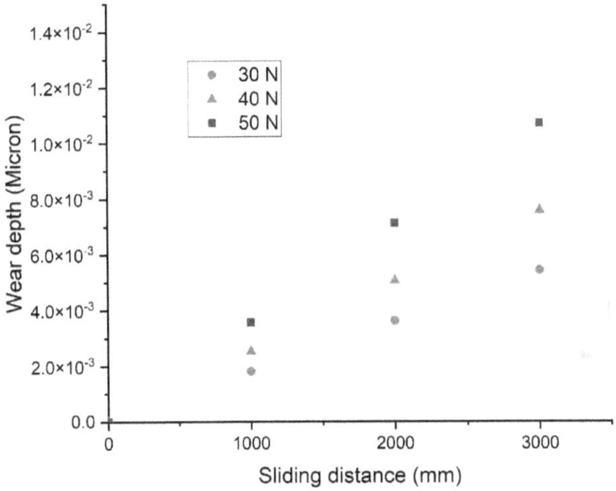

Figure 6.23 Numerical wear results for sprag-on-disc for different wear stages under normal load values of 30 N, 40 N and 50 N.

A comparison of experimental and predicted numerical results for wear depth under three loads for 1.5 hours of sliding is presented in **Figure 6.24.** The present analysis has been carried out for the hardened steel materials in contact for which the wear depth (in microns) has been monitored at particular loading and operating conditions of OWC instead of using the weight/mass loss (in mg) technique as described in relevant literature **[199-200].** The numerical results for wear depth are found to be greater than the experimental results. Such difference is due to the selection of the overestimated wear coefficient measured from the experiments and applied as an input to the wear processor. The local wear evaluated using numerical simulation for specific contact points reflects the nodal values of the wear at each sliding step for one wear cycle. A numerical simulation for one wear cycle consists of several sliding steps for which contact pressure and the relevant local wear are computed at corresponding nodal levels. Out of all these nodal level values, the maximum local wear is considered and compared with the experimental value.

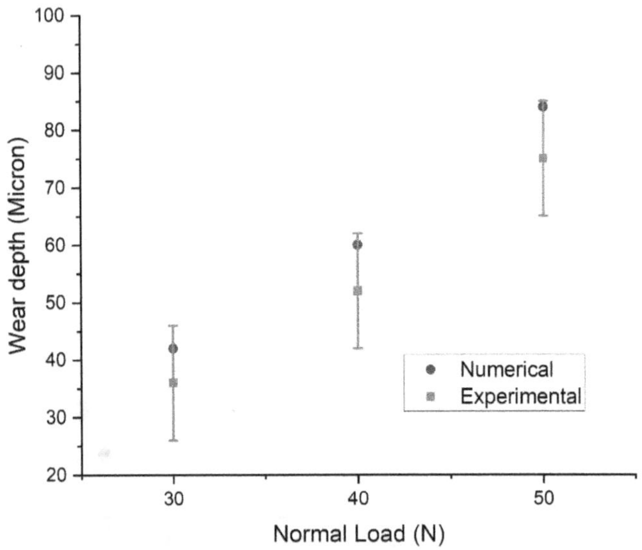

Figure 6.24 Wear depth comparisons between experimental and numerical results for 30 N, 40 N and 50 N normal load running under the same sliding distance.

By expressing the term 'wear at the local scale', the wear is computed at the contact interactions considering individual nodes, whereas 'wear at the global level' refers to the aggregate effects of the wear on the entire model. Since the wear results from the experiments are considered at the global scale and the corresponding wear coefficient values evaluated globally are utilized for numerical simulation at the local scale, the difference in the wear results is observed. The simulation considers the steady-state running periods, while experimental testing includes both steady-state and running-in periods. This results in the selection of an overestimated global wear coefficient, which tends to provide higher wear loss in simulation. The friction coefficient and wear coefficient values from the experiments are derived at a global scale **[201]**, whereas the wear estimation using a numerical approach provides local wear at each node. Therefore, numerical simulation requires local values of this coefficient, which generally vary with time.

Additionally, the presence of debris in actual practice can influence the selection of this coefficient. Moreover, the higher loads introduce plasticity due to higher contact pressures; such phenomena lead to sticking effects and are not considered in the model. As a result, during experiments, wear debris is not removed from contact surfaces in the same amount as in a numerical simulation for higher loads. Due to this reason, the experimental and numerical results differ considerably, but the wear regimes observed in numerical results follow quite a similar trend to experimental results. Additionally, the experimental wear rate is initially higher due to sharp contact asperities, and by considering frictional effects, it is reduced thereafter due to oxide layer formation under temperature rise. Since such thermal aspects are not considered in numerical simulation, the wear depths obtained numerically are greater than the experimental values.

6.6 Overrunning Wear Life of Freewheel Clutch and Life Comparison

During the overrunning function of freewheels, sprags skid on the races at the contact surfaces in direct proportions to the overrunning speed of races. The service life of freewheels is generally specified by the frictional life during the overrunning operation. The duration of the overrunning operation is much greater than the engaging operation of the clutch. Therefore, significant emphasis is placed on assessing the wear life of freewheels. Since the frictional force is very low during overrunning operation, and in the absence of abrasive particles, the failure will not be observed in a short time. Despite this fact, the wear life may vary depending on the lubricating conditions **[202].** In current experimental practice, lubricating conditions are controlled and kept almost similar in each experiment. Since sprags come into contact with the inner race with the help of spring, the drag torque value is influenced by the amount of spring force. The spring force is assumed to be relatively small, and sprags rotate at the same speed as the inner race during overrunning. So, when the inner race overruns, centrifugal force enables sprags to lift off from their positions. As a result, drag torque also varies with respect to the rotational speed of the race. The wear life for three overrunning speeds of 500, 980 and 2000 rpm is calculated based on the experimental and numerical wear results and compared as shown in **Figure 6.25**.

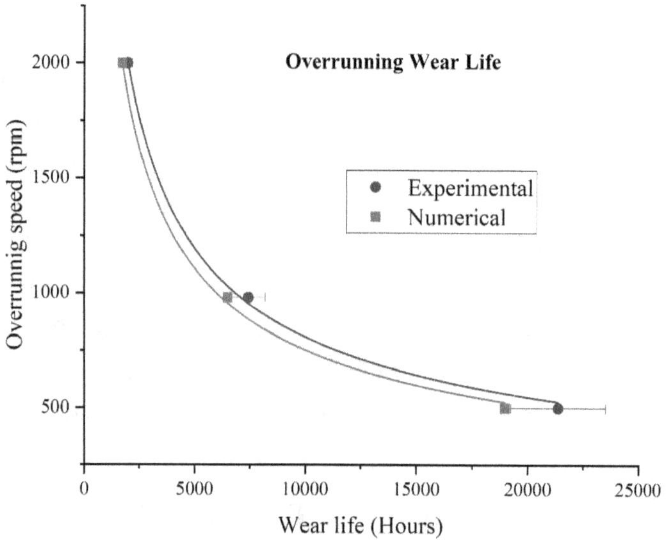

Figure 6.25 Wear life predicted using experimental and numerical results with respect to overrunning speed.

6.7 Summary

The current study aims to determine the wear life of an overrunning clutch from the experimental wear measurements and validate those with the numerical model. In order to analyse the wear behaviour of clutch material, a pin-on-disc tribometer is utilized to conduct wear experiments. Apart from the standard components of a pin-on-disc test rig, a unique clamping device is developed to exhibit required contact conditions similar to the practical operating and loading scenario of the clutch. Based on Archard's wear model for sliding contact, the wear experiments have been carried out using ALTA. The experimental wear amount recorded is utilized to predict wear at operating conditions, and the constants derived experimentally are further utilized to perform numerical simulations based on Archard's wear model. Ultimately, the wear life of the clutch is predicted based on the wear amount with respect to overrunning speed.

Chapter 7

Conclusions and Future Scope

The research work presented in this thesis has thrown light on the fatigue and wear behaviour of the OWCs. As it was critical for the comprehensive understanding, the customized fatigue test rig has been designed and developed for OWCs. An ALTA has been carried out to perform the accelerated tests on the test rig and estimate the experimental fatigue life of OWCs. The computational damage mechanics approach is utilized to estimate the numerical fatigue life of OWCs. The linear elastic kinematic plastic FE-based material model investigates the crack initiation and propagation lives. Initially, a 2-D FE model is prepared, a portion of which is extracted for cyclic analysis. The material microstructure randomness is taken into account by the Voronoi tessellations technique in the RVE contact region for performing the simulation. The material constants of the model are calculated from torsion fatigue data available for case hardening 20MnCr5 steel. Subsequently, damage mechanics-based numerical simulations are performed to obtain crack initiation and propagation lives.

Apart from the contact fatigue observed due to repeated loads, OWC also experiences a large amount of wear during overrunning operations. The experimental wear measurements are executed using a pin-on-disc tribometer with a customized clamping device suitable to hold the contacting sprag instead of the pin with the revolving disc. The wear data are recorded for the specified loading and operating conditions using the concept of accelerated tests, and the experimental wear and frictional characteristics are estimated thereafter based on Archard's wear model. An FE numerical model utilizes the adaptive mesh technique to capture the deformation under load, thereby predicting the numerical wear life using a wear processor.

7.1 Conclusions

The major conclusions of the research study are presented below.

1. An L-P model for fatigue life estimation is applied to clutch samples under study, which defines an inverse power load-life relationship with respect to contact pressure at different torque levels. A fatigue life equation for the OWC specimen is derived in terms of the torque-life relationship that is similar in structure to the original L-P model.

2. An accelerated life test analysis is carried out for the experimental data to develop a load-life relationship, which is found to follow the Weibull distribution as a life characteristic at elevated torque levels with 4-5% reduced life than the L-P model due to the presence of surface traction.
3. The surface damage due to repeated loading conditions is examined through scanning electron microscopy to understand the governing mechanism of contact fatigue in OWC. The failure examination reveals that surface pitting, smearing, and scuffing are the typical forms of wear observed in OWC under fatigue loading and are responsible for the device failure. It is further concluded that such wear scars or surface depressions gradually increase and progressively remove the material responsible for the functionality failure of OWC.
4. A damage evolution law for elastic and elastic-plastic loading in terms of damage state is utilized in the FE model to capture damage progression with micro-plasticity. The numerical model predicts that the fatigue lives are crack propagation dominant due to material plasticity at high contact pressure. It means more than 60% of total lives are consumed in the crack propagation phase. On the other side, the L-P model predicts the lives which are crack initiation dominant, where around 2/3 portion of the total lives are spent in the crack initiation phase. At low pressure, the current model behaves nearly the same as the L-P model, as plasticity is negligible. The numerical fatigue life results show a good correlation with experimental life results within a 10% difference.
5. The micro-hardness tests are performed to investigate the case hardening effect on fatigue lives. It is observed that hardness linearly decreases in the case region with case depth and remains unaffected in the core region. The hardness variation in the case region is incorporated in the simulation model in terms of change in yield strength. The residual stresses induced by case hardening are estimated using XRD and taken into account in the model. The combined effect of residual stress and case hardening depth on fatigue lives is measured. The largest fatigue lives have been obtained for the case depth of 4b, 6b, and 8b for the contact pressure values of 3 GPa, 4 GPa, and 5 GPa, respectively.
6. The wear behaviour of the overrunning clutch is studied using a pin-on-disc tribometer to evaluate wear life. The wear experiments at three accelerated loads

of 30 N, 40 N, and 50 N are carried out, and the respective wear is measured. The wear life is predicted by extrapolating these results backwards using power law fit.

7. The FE analysis confirms the applicability of Archard's wear model as wear depth increases in accordance with the amount of normal load. The applicability of the FE model is checked by comparing the contact problem results with the Hertz solution. The maximum local wear for three loads is computed as 4.53×10^{-08} mm, 6.32×10^{-08} mm and 8.92×10^{-08} mm for contact pressure values of 167.8 MPa, 179 MPa and 214.5 MPa, respectively, using experimentally obtained wear and friction coefficient as inputs. The wear depths predicted from the numerical simulation are found to be approximately 11% higher than the experimental results. It is because the selection of the overestimated global wear coefficient measured from experiments includes both running-in and steady-state testing periods. Due to the short testing period, the calculated global wear coefficient majorly contains running in period. Additionally, a difference in running in periods for accelerated loads, thermal aspects, and adhesion effects at higher loads resulted in different experimental and numerical results. Despite the difference in wear amount, the wear regime profiles are found to be quite similar.
8. The wear experiments are extended for various overrunning speeds. The overrunning wear life for freewheels manufactured from 20MnCr5/EN31 steels running at 500 and 2000 rpm is experimentally derived using a similar approach and compared with numerically predicted wear lives. Also, it is observed that if the freewheel clutches are correctly assembled and installed with adequate lubrication, the wear lives are recorded to achieve even higher values.

7.2 Scope for future work

The current developments in power transmission industries have augmented demand for one-way power transmission products. With the increase in demand, the effective utilization of OWC in terms of useful life is essential. The current work of estimating fatigue life using the proposed test rig and test methodology will facilitate the manufacturers to predict the life cycle of the clutch more accurately. Moreover, the experimentation using such test rig shall help to develop optimum design solutions for improved life and reliability of the clutch.

The following future research links connected to the work presented in this thesis shall provide scope for future research in this domain.

1. By improvising the load capacity of the test rig and introducing automation and control, the evaluation of fatigue lives can be more expedited than the current capacity.
2. The sampling rate of load-unload frequencies can also be increased to further expedite the analysis with the help of asynchronize 3-phase motor (Servo motor) as a drive and load member.
3. Estimating the total service life of the clutch considering the combined operational span of fatigue and wear during engagement and freewheeling operations.

www.ingramcontent.com/pod-product-compliance
Lightning Source LLC
LaVergne TN
LVHW010726061126
851857LV00014B/1362